THE ART (

IAN NORBURY

SCULPTURES IN WOOD

By Ian Norbury

Fox
Chapel Publishing

1970 Broad Street • East Petersburg, PA 17520
www.FoxChapelPublishing.com

Publisher	Alan Giagnocavo
Book Editor	Ayleen Stellhorn
Editorial Assistant	Gretchen Bacon
Cover Design	Jon Deck
Layout Artist	Linda L. Eberly, Eberly Design Inc.

ISBN 1–56523–222–4
Library of Congress Catalog Number 2004106360

To order your copy of this book,
please send check or money order
for the cover price plus $3.50 shipping to:
Fox Chapel Publishing
Book Orders
1970 Broad St.
East Petersburg, PA 17520
1–800–457–9112

Or visit us on the web at **www.FoxChapelPublishing.com**

Printed in China
10 9 8 7 6 5 4 3 2 1

DEDICATION

This book is dedicated to my clients who have bought my work, some of which is illustrated in this book, much of which, unfortunately, cannot be; to all the carvers who have bought my books and to those whom I have met in my courses; also to my family and friends who have supported me for so many years. All these people I thank for giving me the chance to pursue my art.

ACKNOWLEDGMENTS

I wish to express my warm appreciation to Alan Giagnocavo
who conceived of the book and to Ayleen Stellhorn
who curated and shaped this book.

Table of Contents

Beautiful People 3

Chiyonofuji 4
Cromwell 6
Maasai Warrior 8
Bubbles 9
Slave 10
Samburu Girl 12
Slave Girl 13
Africa 14
Girl with Ribbons 16
Girl with Petticoat 18
Humidor 20
The Stripper 22

Jesters 23

Acrobat 25
Because He's Worth It 26
Fastnacht 28
The Impersonator 30
Justice 34
The Idealist 36
Touchstone 40
Hop Frog 43
Reaching for the Moon 46
The Wheel of Fortune 48
Lord of Misrule 50

Harlequins 51

Harlequin Abandoned 52
Harlequin, the Storyteller 56
Harlequin the Politician 58
The Dream Maker 62
Corruption 64
Autumn Leaves 66

Mythology 67

Gaea 68
Aphrodite 70
Andromeda 72
Moros 72
Bride 74
Merlin 76
Merlin II 78

Literature 79

Ariel. 80
Falstaff. 82
A Midsummer Night's Dream. 84
Puck. 88
Ozymandias. 90

Sick Rose 92
Titania 93
Sheba 94
Pierrot 96
Death of Pierrot 98

Fantasies 101

The Alchemist 102
The Astrologer 104
The Dagger 106
The Drummer. 108
Theatre Mask 110
Venice 111
Secrets 112

The Temple. 114
Europhobia 116
Genesis. 118
Tree Spirits 122
The Box of Delights 125
Il Serenissima. 128

Animals 129

Whales. 130
Dolphins 132
Monkey 133
Dominion. 134

Falcon 136
Racehorse 137
Arkle 138
Tiger 139

The Making of the Masterpiece 140

The Early Years 144

Foreword

Ian Norbury was introduced to me when I was looking for someone to do some carvings of birds for a film I was producing. We had a thoroughly enjoyable meeting, shared a glass or two of wine, and Ian lit a pipe. It's been like that now for almost 20 years; the only deviation seems to include good food and strong coffee!

Ian, of course, turned down my offer of work; he was too busy…, an exhibition was planned…, he needed to travel…. The truth was simply that if he had accepted he would have had to compromise, do someone else's bidding, be a hired hand, and that is not what Ian Norbury is about.

Over the years I have commissioned two pieces from Ian and am the lucky owner of a number more, and I know that Ian has to be given the freedom to explore in order to deliver a creation in an organic way. He has a voracious appetite for ideas, he's a great listener and a formidable raconteur, but you can't buy him; feed him information and detail and the result will inevitably excite and inspire.

Ian is, of course, technically brilliant, but technical in terms of art and creation can sometimes be boring, and that he is not. He is an artist of our time, drawing on the past but creating and delivering work both for and of the present, as well as the future, in a totally uncompromising way. He is able to combine the eye of a cartoonist with the mind of a sharp political commentator.

Like all great artists, Ian invites us to look beyond what we actually see, urging us to open our eyes. To really look, perhaps even to glimpse beyond his own horizons.

Unsurprisingly, Ian Norbury is a complex man—idiosyncratic, charismatic, iconoclastic—a great debunker of pomp, and wonderfully politically incorrect, he is a powerful presence and, at times, perhaps satanic. He is also a passionate and caring man, a man of passion. He has the delicate and sensual hands of a trusted lover who encourages us to expose his subjects further. His eye is truthful, he sees the beauty, but he also sees the pain and doubt; that same unerring eye also reminds us of our responsibilities to this earth and of our own mortality.

It is no wonder to me that Ian Norbury should work with wood, a raw material that is so tactile, warm and sensual. It can also be hard and unforgiving, but with Ian this is a true union of man and material, the one complementing the other to the greatest possible effect.

This book reflects that unity, that unique brilliance.

Simon Channing-Williams
Thin Man Films
Imagine Productions, Ltd.
London, England

Whenever I think of Ian Norbury's work, I think not only of his sculptures, but of his books, courses and exhibitions, as an entire package, a kind of Norbury Machine, kept in good running order by Betty, who provides the dedicated promotion Ian needs, leaving him free to perform his magic.

This formidable partnership has introduced a new genre of figurative art in wood. With this extensive work so enthusiastically promoted, a Norbury following naturally developed, out of which arose imitators ambitious to emulate his success and profile.

The British Woodcarving Association was founded as a result of the interest in wood carving that Ian (and Betty) were able to generate, and included many hundreds of "Norbury devotees."

Ian's client base was uniquely cultivated, and the affluence of his patrons allowed him to indulge his talents freely and with confidence. This cannot be repeated, as each individual must find his own way. Still the imitators try to copy, and many wannabes pay homage to Ian in this unconscious manner.

As this book illustrates, the awe in which his work is held is fully deserved. Ian has matched different woods, used fossils, crystals, precious metals and all sorts of combinations as his imagination dictated, and to stunning effect. Some works are representational whilst others are fanciful or surreal.

This type of work was a shock to many traditionalists who remain stuck in the groove of classical decorative woodcarving and slaves to tradition. The "fine art world," to my knowledge, never embraced Ian's work. This is not surprising in a climate where unmade beds or excrement in a tin wins an award. Nevertheless, it was a breath of fresh air. Bold, even audacious, innovative, and sometimes controversial, his work demanded attention. For the more intellectual, Ian has injected narratives to accompany his pieces, including quotations, poems and philosophical ideas to support (and possibly explain) some of his less obvious creations.

I am not a lover of Jesters or Harlequins, but the execution of the diamond patterned inlays and the detailed accessories compel admiration, especially when one is aware of what it requires to achieve this in practice.

No, the Norbury Machine is unique and will always hold a special place, perhaps even a special field, for it does not lend itself easily to any existing category. I leave that to the academics. All I know is that his work is as dynamic as it is prolific; it is executed with an obsessive quest for perfection, which demands agonizing patience, and the results are as you see—inspirational wizardry!

Ray Gonzalez
Chardleigh House
Somerset, England

One is hard pressed to describe and appraise the work of Ian Norbury without sounding sycophantish. In sculpture generally, and in the carving world specifically, it is the norm to see the human form rendered as "still life." To contrast that with creations such as *Harlequin Abandoned* brings positive delight. Beyond the fine workmanship and seldom matched exploitation of the beauty of wood, Ian's work is characterized by the inviting and intriguing balance of opposites. The static and subdued detailed facets act in concert with elements subtly nuanced and articulated with frugal economy; the raw and evocative play against the passively sublime. His creations demand one's attention and admiration.

He is furthermore a brave soul, this Ian. Many a mediocre artist indulges "Multimedia" as an easy fix, a means of achieving originality simply through new combinations of old tricks from separate spheres. The results are often so void of esthetic merit that they can hardly be called advances. The pitfalls snare them, and their works smack more of novelty than of novel. Against the principle medium, the added ingredients appear intrusive and extraneous and seem contrived and supplementary rather than complementary.

Ian's skill, intelligence and wit would place his works at the top of his genre if he chose to render them entirely from one wood. Yet he chooses to entertain the challenge and does it well, increasingly well at that, particularly where he is himself masterful in the added medium. It is exciting to anticipate where his creativity and explorations might next lead him.

Ian Norbury first came to my attention via the back cover of *Fine Woodworking* magazine, an article on his *Alice in Wonderland* pieces. It was immediately apparent that this was an individual of stellar potential. When apprised of the amount of energy put into his books and travel and teaching,

Foreword

I worried that his own development would plateau as has that of many others sucked into the seminar trap. Gratefully, it seems to further energize and inspire him instead. Amazing! (And it's suspected that the supportive efforts of his Betty deserve much credit!)

It's often said that familiarity breeds contempt. True enough, if it's exposure to the mediocre and banal. But when the acquaintance is to skill and ingenuity creatively manifested, there is the promise of increasing fascination and admiration with time. This volume will leave its examiners desiring to better know its author. Artists of Ian's standing acquire a mythic stature, and it takes time and opportunity to sort out fact from fiction. If you have the chance to enroll in one of his seminars, seize the opportunity. Otherwise, follow the Biblical injunctive to "Know him by his works." You'll be neither misled nor disappointed.

Fred Cogelow
Willmar, Minnesota

I first met Ian and Betty Norbury in the Spring of 1982 and in a rather curious fashion – driving past his gallery in Painswick Road in Cheltenham I noticed his carving of The White Knight (which subsequently became his particular logo) and remarked to my driver on the superb detail. He responded with the remark that it was probably plastic, as no woodcarver could achieve that intricacy of detail! A visit (my first) the following day proved him to be very wrong! That chance encounter turned out to be the genesis of a lasting friendship: The Norburys (for they are an entity) were true friends for all seasons. Consequent on this close acquaintance came the privilege of seeing Ian at work: his methods, his skill and his unique insight into the psyche of his manifold subjects; particularly those that could be traced to Celtic origins. Being myself a Scot of Highland descent with a life long interest in Celtic legend, music, history and art, I was in a position to pass some degree of qualified judgment. To quote an old Highland expression, "...he has the very spirit of the thing."

I have always felt that Ian surpasses many of the great woodcarvers of the past in the sheer breadth and versatility of his work – ranging from his depiction of the underground temple at Tarxien in Malta to glorious heraldic beasts; all carved with meticulous attention to detail. To describe Ian as multi-talented would be an understatement, for the man possesses true genius.

Of Ian the man, I would say this: Ian inhabits two worlds – art and commerce and it is sad to state that the second of these has frequently an adverse effect on the first. In Ian's case the reverse applies – art always overrides commerce. His integrity and standards of conduct towards his patrons is of the highest.

Behind every successful man there exists an often unseen motivating force, to organise and where necessary to inspire – Betty Norbury provides this for Ian.

Over the years I have had the opportunity to buy several pieces of Ian's work and occasionally to commission others: they give continued pride and pleasure to me and to people who have seen them. This beautiful book goes some way to revealing the span of Ian's talents, and the great humanity of his vocation. I wish him and his lovely wife continued good fortune and success in all they do.

Donald Cameron Wilson
Lord of Lanercost

I first met Ian Norbury in 1970 when he was twenty-two years old. At that time he had been living in Cyprus for four years, happily painting pictures for tourists, portraits for the locals and murals for nightclubs. In 1971 we returned to England to marry, and Ian began painting racehorses and foxhunting scenes, very commercial subjects, but we had a family to support.

After a disappointing exhibition in 1974, he decided to become an art teacher and studied sculpture at St. Paul's College, Cheltenham, for three years where his interest in woodcarving began. After attaining his degree, he bought a house, gallery and studio in Cheltenham and began carrying out carved restoration work for the antique trade. After a couple of years, he changed to sculptural work and in 1983 published his first book, *Techniques of Creative Woodcarving*, followed by two more in 1985 and 1987. He also wrote extensively for woodworking magazines, judged competitions, gave courses in England and Switzerland and held an annual one-man exhibition. Some of these years were difficult, but our children and I supported him one hundred percent.

By the late eighties Ian's exhibitions were very successful indeed, and he had become an international figure in the woodcarving world, judging at the Canadian National Exhibition in

Toronto, giving lecture tours in Europe, America, Australia and New Zealand and becoming chairman of the British Woodcarvers Association.

His success continued: publishing more books; holding annual exhibitions; and teaching in Iceland and Ireland, where he set up a second home and studio in 1996 while maintaining his original home, gallery and studio in Cheltenham.

Most of his work is speculative, and although he undertakes commissions if they are interesting, his refusal to carve anything that he does not want to carve, which has been a source of frustration to me, his bank manager and accountant, is undiminished since he sold his first painting at the age of fifteen.

As a leading entrepreneur in the woodworking arts field for over twenty years, I know that Ian is very successful compared with most of his peers, both financially and in achieving what he wanted to do. When we opened the studio and gallery in the early eighties, friends laughed at us—people were not exactly scouring the country to buy wood

Ian's gallery, the White Knight Gallery, and studio in Cheltenham, England.

Ian's studio in Ireland is nestled in the rural Irish countryside.

Ian's ability as a sculptor is rivalled by his success as a teacher. Many of his students have been astonished at what they are able to produce under his direction and how much can be achieved in a short time using his methods. His innovations in techniques and concepts revolutionized woodcarving and, to a large extent, created the great popularity it enjoys today. He has always been willing to share any knowledge he has and to help anyone with a genuine interest.

In the future Ian will continue to push the boundaries of woodcarving, perhaps in directions which will dismay the traditionalists, enrage the politically correct and delight his fans.

Betty Norbury
Gallery Director

Organiser of Celebration of
Craftsmanship and Design

Author of
Furniture for 21st Century
Promotion and Marketing for Crafts
Fine Craftsmanship in Wood (U.S.A.)
British Craftsmanship in Wood (U.K.)

sculptures. Ian's attitude was that if you make something very special, somebody will want it. This is the secret of his success. He makes things that astonish people by their virtuosity of execution, intrigue people by their enigmatic mystery or enchant people by their beauty. I believe he learned this from Salvador Dali, an artist whose work is totally baffling to most of his audience, and yet they find the combination of Dali's technical expertise and mystery compelling. It is for this reason Ian has pursued excellence in technique, because when you try to create an illusion, nothing must detract from it—no technical fault must break the spell.

His sources of inspiration are very wide, as anyone who has seen his library can vouch for, and are from a widely read knowledge of art, history, folklore, philosophy, psychology and a lot more. Like a magpie, he picks out the jewels that attract him. Probably the artists who have given him most are Barlach, Dali, Magritte, Fuchs, Jonssen and, most of all, the Mexican surrealist Remedios Varro.

Above all, he is steeped in European cultural history and has an overwhelming sense of the fatalistic repetition of history and the inevitability of the archetypal errors of humanity endlessly repeating themselves. It amuses him to illustrate the folly of mankind in his sculptures with the advantage that he has empirical knowledge of most of them.

Ian at work on a new piece in his Cheltenham studio.

Beautiful People

A lthough most of my carvings originate from an idea which has then been researched and built on, some are more spontaneous, simply arising from an image of a particular pose, such as *The Stripper*, something read in a book, such as the *Maasai Warrior*, or the possibilities of a particular piece of wood, such as the *Girl with a Petticoat*. Others are commissions from clients and occasionally just a relaxation from inlaying diamonds into a harlequin.

Lime, purpleheart, boxwood and sycamore
27 inches tall

Chiyonofuji

When I was commissioned to make a carving of a Sumo, my first thought was of the short, violent struggle between two enormous men. However, I felt there were two objections to carving this scenario. Firstly, it is not really feasible to have one Sumo fighting; you need two. Secondly, the carving was to be installed in a peaceful Japanese-style gallery, overlooking a Japanese-style garden. A violent type of figure seemed inappropriate. I therefore portrayed the Yokozuna Chiyonofuji performing the Dohya ceremony, with his arms outstretched to show that he is unarmed and comes in peace.

Chiyonofuji was one of the great Sumos, smaller than most but immensely strong. He became a Sumo in 1981 and remained so, increasing in power and success, until 1988, when after 53 consecutive victories in that year, he was defeated by Onokini on the final day of the November tournament.

I feel that this is a most successful piece with a good balance of plain wood, detail and colour. It is a very friendly and welcoming object in the clients' house.

Detail, back view: Although a small man, Chiyonofuji's immense strength was made obvious when he picked up a three hundred pound opponent by the belt and threw him out of the ring.

Cromwell

Oliver Cromwell, 1599–1658, Lord Protector, seems to convey the idea of a puritanical moraliser whose rather boring supporters managed to overthrow the dashing cavaliers of Charles I. He was, in fact, a brilliant cavalry commander, a military genius, and the perpetrator of the massacres of Drogheda and Wexford, for which his name is still hated in Ireland. This carving was commissioned by the owner of a superb Cotswold stone and half-timbered mansion called Cromwell Court.

I positioned the horse and rider in this typical heroic pose, similar to David's portrait of Napoleon, because I feel it not only conveys the image of vigorous war-like activity, but the rearward gaze seems to indicate some purpose or direction and suggests distance and space. Cromwell and his horse are dressed in the uniform and accoutrements of an officer of the Ironsides, the cavalry corps designed and created by him.

Lime
36 inches tall

Maasai Warrior

The problems in Sub-Saharan Africa have been so horrendous during my lifetime that most people have given up even thinking about it. We are appalled by a few tragic deaths in Europe or America while Africans die in the millions. The extraordinary brutality of the tribal conflicts has destroyed forever the notion of the "noble savage."

One tribe that seems to have emerged unsullied is the Maasai of Kenya. These people have basically rejected Western culture and have maintained their traditional life, herding cattle and living in villages. The tall, athletic young men, renowned for long distance running, kill lions with their bare hands to prove their manhood.

Walnut, silver, bone and marble
20 inches tall

Lime, quartz and marble
15 inches tall

Bubbles

Many of my pieces are very complex and take five to six weeks of concentrated, painstaking work—always with the nagging possibility that at the end of the day any given piece might not work. After such an episode it is almost like a holiday to carve something comparatively plain and simple, like a nice little nude from easy-to-carve limewood, that I can be 99% certain will work. Bubbles is such a piece and, I think, one of my best nudes, delicately and subtly enhanced by the quartz crystal balls.

Slave

Looking back, I seem to have carved a lot of people in chains; no doubt somebody could explain that. Many artists have used slavery as a subject, such as the American sculptor Hiram Powers, whose Greek Slave horrified people in 1846, not because the woman was a slave, but because she was white.

Walnut and bone
16 inches tall

Samburu Girl

The Samburu are closely related to the Maasai but live in an even more remote area of Kenya and have remained relatively untouched by outside influences. The young women are very beautiful.

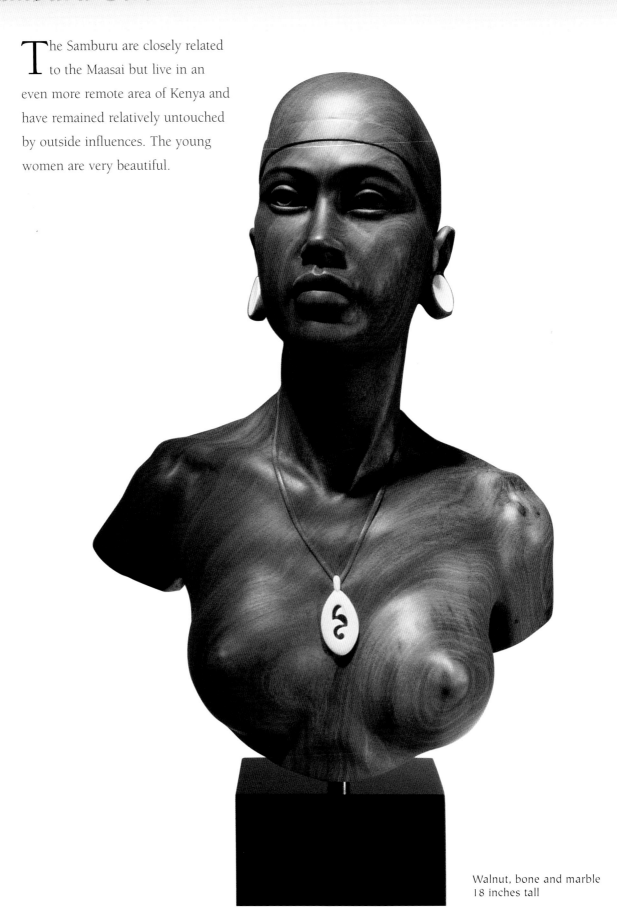

Walnut, bone and marble
18 inches tall

Slave Girl

This was my first female slave, made from a large block of walnut. It was also the first time I used metal on a carving. I paid a jeweler to make proper, hinged manacles, mainly because I was tired of delicate pieces of wood being broken. I was so pleased with the result that I used metal more and more afterwards.

Walnut and silver
16 inches tall

Africa

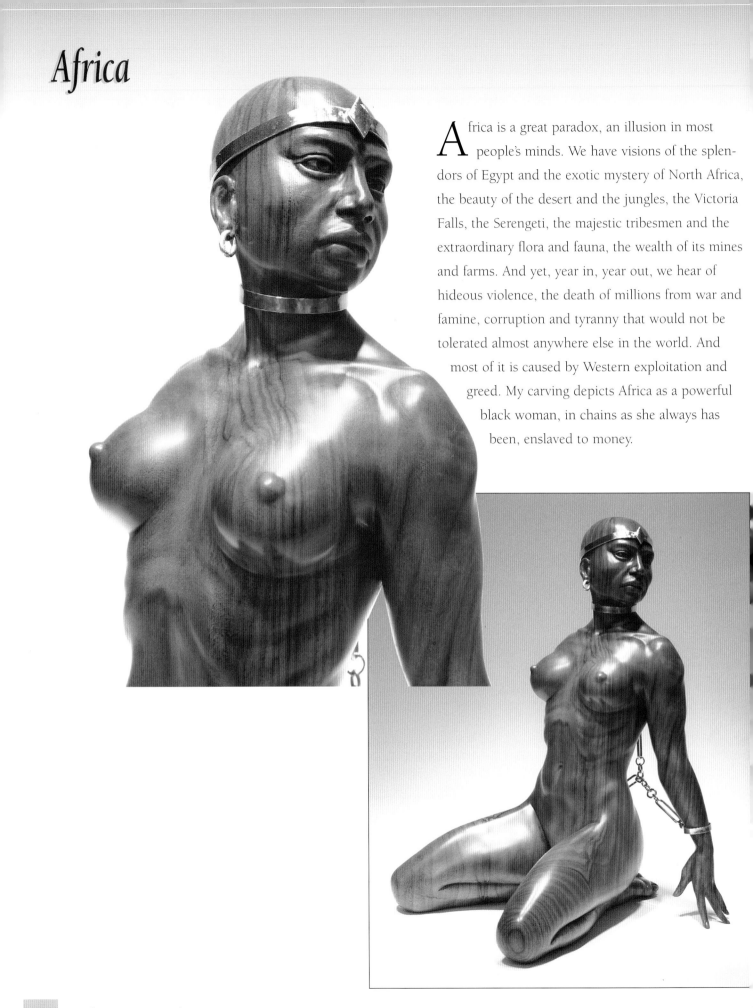

Africa is a great paradox, an illusion in most people's minds. We have visions of the splendors of Egypt and the exotic mystery of North Africa, the beauty of the desert and the jungles, the Victoria Falls, the Serengeti, the majestic tribesmen and the extraordinary flora and fauna, the wealth of its mines and farms. And yet, year in, year out, we hear of hideous violence, the death of millions from war and famine, corruption and tyranny that would not be tolerated almost anywhere else in the world. And most of it is caused by Western exploitation and greed. My carving depicts Africa as a powerful black woman, in chains as she always has been, enslaved to money.

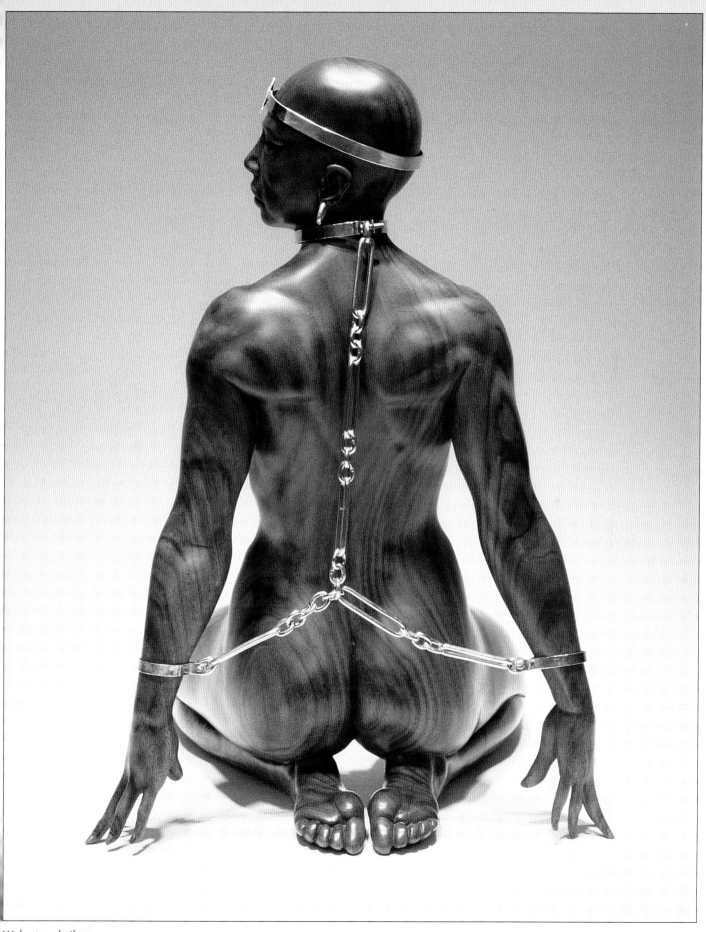

Walnut and silver
16 inches tall

Girl with Ribbons

Forty years ago, elms were one of the commonest trees in England. The timber had no great value despite the huge size of the trees and the strength of the wood. It was used for coffins, the seats of spindle-backed chairs and props in coal mines. When Dutch elm disease exterminated the elms, it was a great loss to the beauty of the countryside, but it did put elm on the map as a desirable timber.

When I bought my cottage in Ireland, there was a dead elm in my garden, about 15 inches in diameter. From the trunk I made two nudes, each four feet high: Girl with Ribbons, shown here, and Girl with Petticoat, shown on the following page.

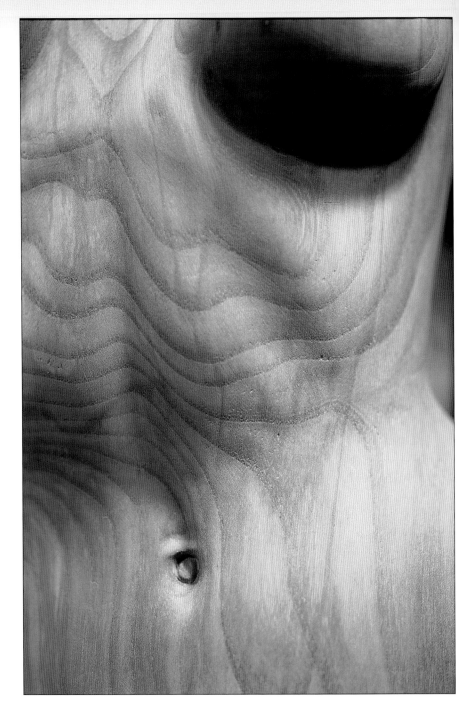

Elm and copper
48 inches tall

Girl with Petticoat

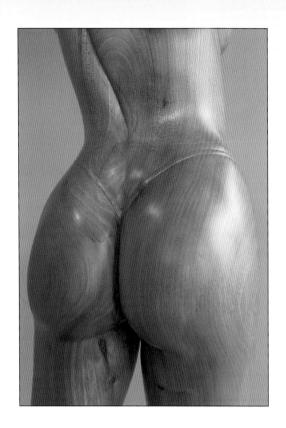

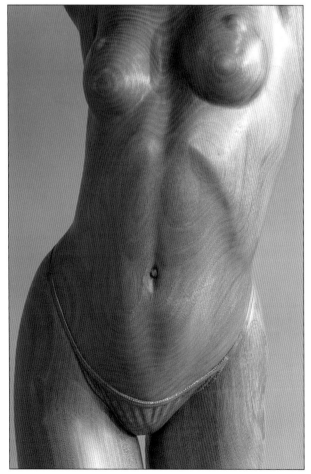

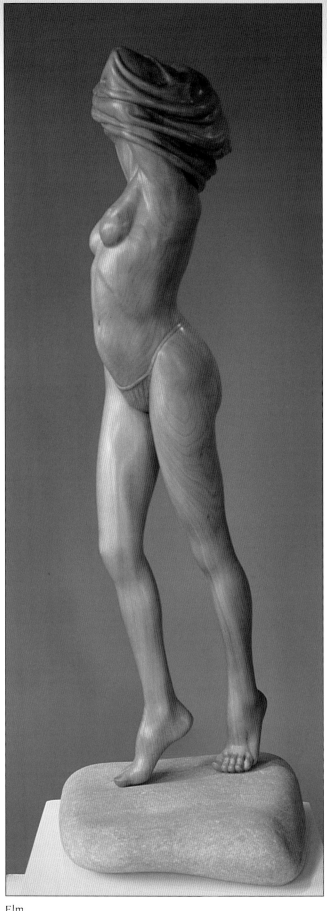

Elm
48 inches tall

Humidor

Cigar smoking is a growing pastime in Britain, America and many other countries. Huge amounts of money are spent on these fragile objects, and much care and devotion lavished on keeping them perfect condition for burning. Humidors range from cheap little boxes fitted with a wet sponge to entire air-conditioned rooms.

I made the carving on this humidor at the time of the Monica Lewinsky scandal in America, mainly as an experiment in making stockings. The whole design is based loosely on the Art Nouveau style as exemplified by Alphonse Mucha in his advertisements for cigarettes with beautiful, swirling smoke. The girl has that sultry, nineteenth-century Parisian look so often seen in old photographs.

Lime, ebony, walnut, purpleheart, maple, baroque pearl, amboyna, bolivar and cedar of Lebanon; humidor by Andrew Varah
18 inches long

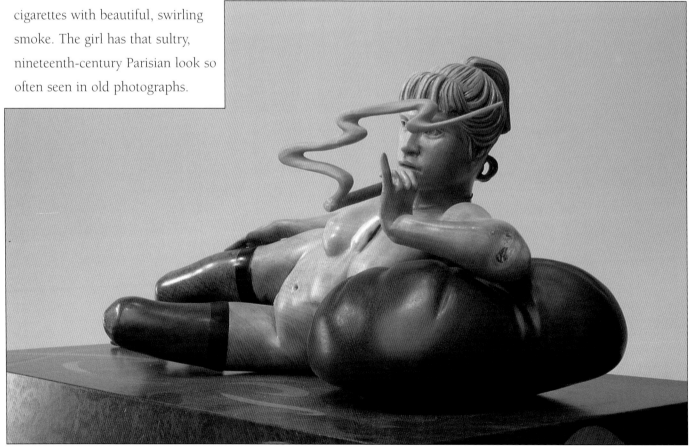

The Stripper

It is a popular conception that strippers are actually unhappy and sad when they are not on stage appearing to be having a great time. I doubt this is true of that profession any more than any other, but in this carving I have tried to portray the poignancy of that image.

The lime figure is seated on a bar stool made of oxidized brass and ebony, and she is holding a scarf made of pink ivory wood. The base is in the form of an ebony veneered box with an inlaid "spotlight" of sycamore and ebony shadows running away from the legs of the stool. I consider this one of my best portrayals of the female form.

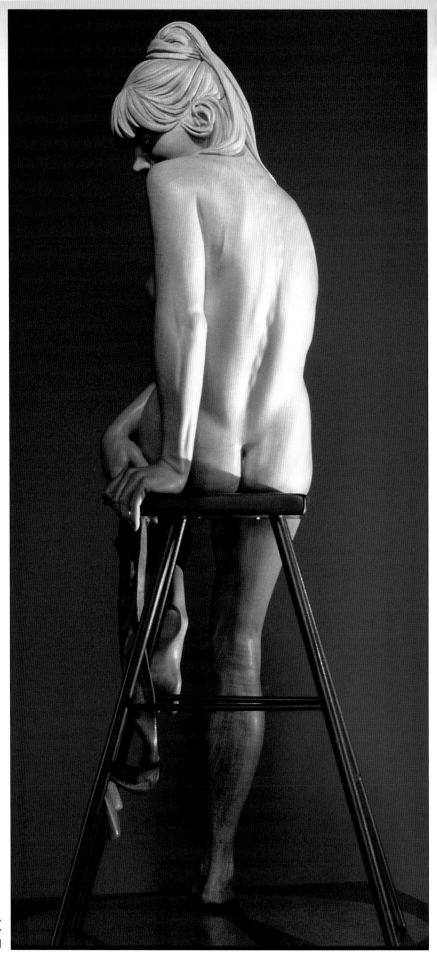

Lime, pink ivory wood,
brass and ebony
24 inches tall

Jesters

T his fellow's wise enough to play the fool. And to do that well, craves a kind of wit" 'Twelfth Night' William Shakespeare

The jester (from the Old English jest, an exploit) was more commonly known as a fool or buffoon. The word fool derives from follies, meaning a pair of bellows, a windbag or scrotum. Buffoon comes from buffare, to puff. The Fool in medieval times was as much an institution in the courts of kings, princes and aristocrats as the chamberlains, court officials and minstrels were. For several hundreds of years they were an integral part of aristocratic life, occupying an ambivalent place in people's minds. Primarily they were a source of amusement and were abused in ways that would horrify us today. Often they were in fact mentally unbalanced or retarded and hideously deformed. Indeed the countryside was scoured to find such people.

In contrast to this callous behaviour there was a belief that the utterances of idiots were divinely inspired and revealed some truth, but in a garbled fashion, rather as Freud would have us believe of dreams. Very often the court fool was on the most intimate terms with his master and could say things to him that would have resulted in disaster for a courtier. Many became wealthy and famous. Of course some were not natural fools, but were very clever people pretending to be mad. Some even wrote their autobiographies.

The Fool's costume was traditionally the cap, bells and motley, a ragged assortment of garments. The cap or hood had long dangling "ass's ears" with bells on the ends as can be clearly seen in Durer's engravings illustrating Sebastian Brandt's book *Der Narrenschiff* (The Ship of Fools), 1494. Often his cap has a cock's comb or a cockerel's head as a crest, a symbol of stupidity but also virility. Cock-crested figures can still be seen in Fastnacht processions and bizarre cock-criers crowed the hours in the English court well into the 18th century.

Of course the costumes changed, and in many illustrations the jester is shown in a floor length, full-skirted cloak covered in bells. The Fool also carried a mace or bauble, derived originally from a small club, that gradually developed into a wooden rod with a carved replica of the fools head with its tongue sticking out on one end and a pig's bladder representing a phallus hanging from the lower end. With this bauble the jester would perform comical and obscene acts to amuse and insult his audience. He would also carry on conversations with it rather like a ventriloquist. Thus the bauble might outrageously slander the king, and the Fool would vigorously defend him, allowing the insults to be made but deflecting blame from himself.

The official post of court jester ceased to be a significant institution with the death of Charles I but lingered on here and there. In Berkeley Churchyard there is an epitaph to the Duke of Suffolk's Fool written in 1728, and in odd parts of Europe they continued into the late 19th century.

So why make carvings of jesters? Apart from the interesting things one can do with the actual figures and costumes, the jester still fulfils his role, at least in Europe, where he is embedded in the culture, of providing a tongue-in-cheek reflection of the folly around us, whether it is in Shakespeare, Charlie Chaplin, Danny Kaye or Basil Fawlty banging his head on the desk in frustration. Whatever century, Erasmus' words in *The Praise of Folly* (1509) still ring a bell. "Now what else is the whole life of mortals but a sort of comedy, in which the various actors, disguised by various costumes and masks, walk on and play each one his part, until the manager waves them off the stage? Moreover, this manager frequently bids the same actor go back in a different costume, so that he who has but lately played the king in scarlet now acts the flunky in patched clothes. Thus all things are presented by shadows; yet this play is put on in no other way."

Acrobat

I have always been led astray by the pursuit of technical difficulty in the same way that men pursue beautiful women. The lure of trying to make a piece of wood do something it does not want to, I find hard to resist. I am aided and abetted in this by a large proportion of the public who greatly admire technical dexterity and are prepared to pay for it. A huge amount of the world's art, which is held up as the best, is primarily very clever—very often the inspiration behind it is only dimly perceived.

The Acrobat is one of my early attempts at being clever and is reasonably successful in that the flying walnut ribbons and the thin lime of the paper hoop do give it a certain lightness and movement, but it is really an exercise in technique.

Walnut and lime
16 inches tall

Because He's Worth It

This is one of a series of carvings I am doing on the theme "Seven Deadly Sins for the 21st Century." The original sins devised by Pope Gregory the Great in the 6th century are socially past their "sell-by date."

"Avarice" is an admirable quality now, and with the advent of the lottery, it is a national institution. "Sloth," personified by the television addict laid on the sofa with a six-pack, is the stock-in-trade of the companies who broadcast the never-ending stream of rubbish that fills the airwaves.

Here we see my Fool admiring himself in a hand mirror in the shape of a heart. His costume is decorated with narcissi flowers, the symbol of self-love. But a slight frown furrows his brow: "Can this be a wrinkle?" In the mirror, his reflection is the face of Michelangelo's statue, David, an icon of perfect manly beauty, which is how the Fool sees himself. David is also looking suitably concerned about the offending blemish. On the reverse of the mirror, a devil's face laughs at the Fool's self-delusion; the devil lives behind the eyes.

Of course today there is no harm in this—pandering to vanity is a billion dollar industry. The Fool's sin is that he is obese. In today's world that is unforgivable!!

Detail, Face: The Fool's pudgy, effeminate fingers trace the line of a possible crow's foot, marring the perfection of his face.

Detail, Costume: In Greek mythology, Narcissus fell in love with his reflection in a pool, but fell in and drowned. He became the flower of that name, the symbol of self-love.

American cherry
24 inches tall

Fastnacht

In late February, when the Carnival is taking place in Venice and Mardi Gras in New Orleans and elsewhere, north of the Alps in Switzerland they are celebrating Fastnacht. While the Carnival in Venice has its origins in 18th century culture with its elegant costumes and beautiful masks, Fastnacht goes back much further to pagan times when the people drove away the spirits of winter by making as much noise as possible. For four or five days the German Swiss of Lucern put on bizarre costumes and masks, which they have spent all year making, and roam the narrow streets and squares of the old city, drinking, eating and making noise. Some 80 brass bands patrol the town all playing different music, mostly badly. The costumes tend to be far more Gothic than in Venice, demons and monsters predominating, although many are political satires or simply funny. At midnight on the last day it all disappears as if by magic, and normal service is resumed.

My carving shows one such reveller in the costume of a jester beating his tambourine for all he is worth.

Walnut, bone, Mexican rosewood and sycamore
14 inches tall

The Impersonator

Walnut, pear and mixed media
30 inches tall

There is a wonderful cartoon showing a circus clown standing in front of a mirror. The reflection shows a city executive with a suit and bowler hat. The two regard each other with the same wistful expressions. Does the clown wish he were a business man or the business man wish he were a clown? Or is there a business man inside every clown and a clown inside every business man? Maybe the point is that the clown actually sees himself as normal. And then, is not the business man simply putting on his own mask when he dons his city uniform? Is this not the persona he assumes for his meetings, which he will shed when he gets back home to his family, only to adopt another when he goes to the local pub or out to see his mistress?

The Impersonator is a fool, dressed as a court jester in cap and bells. He is festooned with masks, the many personas for his various performances, all of which are interchangeable. He is standing on a base with four windows painted with scenes from his life. The front one shows him sitting on a throne in an empty palace, a lonely King of Fools.

Detail mace: A small fool tops the mace held behind the jesters back.

Detail face: The anxious, vulnerable face of a man who no longer knows who he is, peers from behind the mask.

Justice

"In England, Justice is open to all—like the Ritz Hotel." –Sir James Matthews, an Irish Judge.

"The Law is an ass," they say, and so it would seem fitting that the figure of Justice should be wearing the "ass's ears," cap and bell of the jester or court fool.

Parodying the statue of Justice on the Old Bailey in London, the sword is replaced by the fool's mace with its bladder and topped by a head of Mr. Punch blowing raspberries at the law. In his other hand Justice holds the scales; in the pure silver pan, the pearls of truth weigh heavily against the block of gold in the base metal pan. However, Justice has spotted this and extends his little finger to tip the scales in favour of the gold. The loose hanging sleeve is embroidered with symbols of those sections of society perceived as being immune to justice: politicians, the famous, the wealthy….

This overstuffed dissolute pillar of the law bestrides the world, a ball of labradorite or wizard stone, atop a pile of decaying legal tomes and manuscripts.

Detail head and shoulders: Smiling smugly "blind Justice" peeps from under his blindfold to see which way things are going. Curiously the concept of blind justice, in which the blindfold symbolises impartiality, is a recent innovation. In olden days the blindfold symbolised stupidity and blindness to knowledge.

Walnut, lime, boxwood, pink ivory wood, labradorite, silver, gold, copper, pearl
35 inches tall

The Idealist

Idealists are always a bit worrying: If we agree with them they are single-minded, and if we do not they are bigots. Nevertheless, the true idealist makes you feel a bit uncomfortable, because you do not keep to what ideals you may have in the way that he does.

My Idealist travels through life on his precarious three-wheeler, which has no means of steering, but his eye is fixed immovably on a distant and unattainable point of perfection that he can see through his telescope. He is dressed as a Fool, his garments ragged and his body emaciated. He has no thought for appearances, his mind being on higher things.

The base is painted with a view of the smoky nebulas of deep space, the lofty realms he inhabits. The quotation around it, from Charles Dickens, reads: "Softly sleeps the calm ideal in the whispering chambers of imagination."

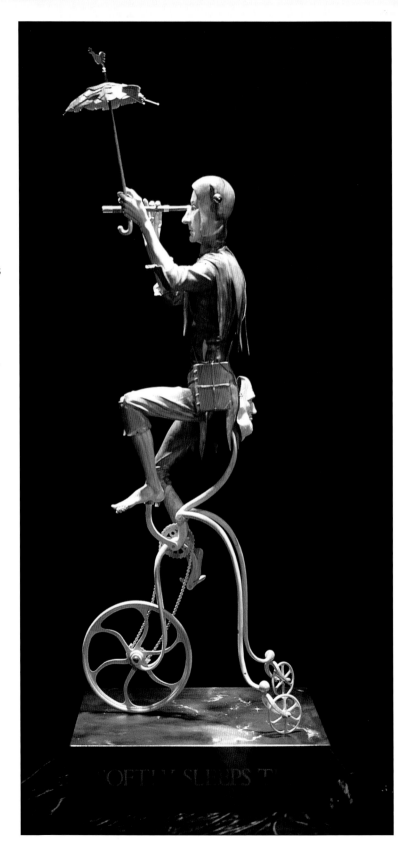

Detail head: The lean and hungry face of the Idealist stares through his brass and thuya root burr telescope. His buttons are made from various semi–precious stones believed to possess useful virtues.

Lime, boxwood, walnut, yew, copper, sycamore, brass, thuya root burr, ebony and semi-precious stones
30 inches tall

Detail mask: Vital to the Idealist is his mask, which he is ever ready to hide behind. It takes the form of a bearded philosopher in the style of Plato, Socrates, Leonardo, Darwin and Marx. All great thinkers have beards.

Detail book: This is the Idealist's book of his writings, directed against his enemies. It is appropriately made of yew, a poisonous timber traditionally used for making bows and arrows.

Detail umbrella: The Idealist protects himself from criticism by supportive quotations from learned men. His umbrella therefore consists of pages from books topped by a weathercock—just to see which way the wind is blowing.

Detail leg: The Idealist's sinewy legs propel him relentlessly on his way. The fragile looking tricycle is actually remarkably strong, as the structural components all follow the natural grain of the boxwood.

Touchstone

Touchstone is the court jester in Shakespeare's comedy *As You Like It*. His caustic wit and penetrating comments on human weakness and folly give him an aura of jaded worldliness, creating one of the Bard's most unforgettable characters. He is shown elegantly posing in the full regalia of the medieval Fool.

Walnut, sycamore, gold, malachite, red jasper and cocobolo. 26 inches tall

Detail upper body: The hood is decorated with a cock's comb. The cockerel was seen both as a symbol of virility and courage but also as vain and stupid because of its constant strutting and crowing. The hood also has the traditional ass's ears.

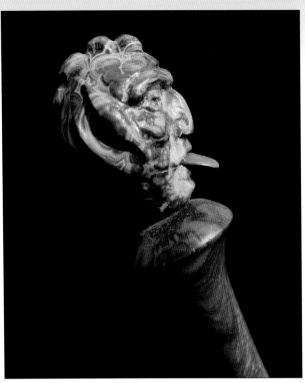

Detail mace: The head on the mace is carved from malachite with a tongue of red jasper on a shaft of cocobolo.

Detail sleeve: These large sleeves were high fashion in the 14th century. Clothes were put on in pieces and tied together with laces.

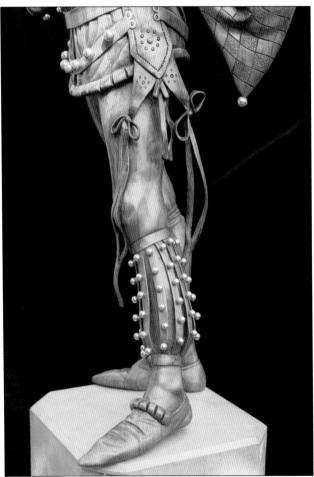

Detail leg: The straps of the satchel and the bells on the leg present some very difficult carving. Bells of this type are still in regular use by Morris dancers in Europe.

Detail left sleeve: The slashing of the sleeves and body of tunics was a popular fashion. The top layer of cloth of one colour was cut to reveal contrasting colours below.

Hop Frog

"Hop Frog" is the title of a short story by the American writer Edgar Allen Poe. In Poe's story, Hop Frog is the crippled dwarf jester of a cruel unnamed king who so abused Hop Frog's lady friend, Tripetta, that he decides to take his revenge at a masquerade in the court. He persuades the king and his seven councillors to dress up as "ourang-outangs" (*sic*) in order to frighten the women and duly covers them in black tar and chains them together. He also arranges for the chandelier to be removed from its chain. At the height of the festivities, he hooks the chandelier chain to the "ourang-outangs" (*sic*), hauls them up and sets fire to them, escaping through the skylight.

Hop Frog is depicted here at his moment of triumph. He is made from a massive block of walnut and hangs from a high ceiling on a chain.

Walnut and copper
36 inches tall

Detail feet: On this large scale, the tooled finish brings out the beautiful grain of the walnut.

Detail torch: The flames of the torch are made from heat-treated copper sheet.

Detail head: The mask is made from sheet copper, which has been hammered into shape over the carved face underneath.

Reaching for the Moon

Fortune was depicted in many different ways, usually as a woman, sometimes with wings and carrying various implements. Invariably the figure stands on a ball, which represents instability and perhaps the world. Sometimes the figure was of a youth with a long lock of hair hanging from the front of his head—for grasping Fortune by the forelock—and holding a balance resting on the edge of a knife.

This piece is loosely based on that image, but Fortune is replaced by a Fool standing precariously on one leg, reaching for the moon, which he is never likely to reach since it is attached to his own head, his dream being an illusion created in his own mind. In his other hand he holds a knife on which is balanced a set of scales, in which the Fool's life is contained. A pink ivory wood heart represents his spiritual life and an ebony padlock symbolises his material life.

Walnut, silver, copper, brass, pink ivory
wood, ebony and leopard skin jasper
30 inches tall

Detail close up: Although he risks everything he has reaching for the Moon, he does have a silver wheel of fortune pinned to his chest for luck. As well as having ass's ears on his hood, typical of the medieval jester, he also has a cock's head on the top. The long ribbons on the sleeve and ridiculous shoes were high fashion in the late 14th century.

The Wheel of Fortune

The wheel of fortune is a concept dating back to antiquity when it symbolised ephemeral happiness and the fortuity of the moment. It became very popular in medieval art where it is frequently encountered as a wheel being turned by Fortuna, with figures clinging to it. It symbolises change of fortune and the permanent fluctuation of all that exists.

The wheel and the well (The well represents the entrance to the Underworld but also has wish-fulfilling properties.) are made from walnut. Standing on the well-head is the beautiful Goddess Fortuna embodying the seductive attractions of the world of commerce. She is carved from lime, traditionally seen as a symbol of femininity but believed to attract disease when touched. Fortuna controls the wheel, and she holds the Thread of Destiny of those on it. Blindfolded, she turns the wheel at random and on a whim, cuts the threads with her silver knife.

On the wheel are three figures dressed as Fools, symbolising the folly of the pursuit of material wealth. On the left, an aggressive young man is vigorously scrambling upwards. He is made from mulberry, traditionally representing the rising sun. He is decorated with silver and copper, not yet having acquired wealth, but is lead on by the gold bauble dangling before his eyes. His buttons are garnet, which was believed to warn its wearer of danger by changing colour. His bells are lapis lazuli to inspire courage and confidence. His copper mask symbolises deception, so often the path to riches.

On top of the wheel is the golden boy wearing the crown. He is carved from boxwood, symbolising immortality, and is decorated with gold. Money is falling from the sky into his hand but falling out again just as quickly. His mace is a folded umbrella suggesting the city businessman, made from olive for victory and lignum vitae for long life. He smiles, smugly oblivious to the sight of his predecessor plunging into the abyss.

The third figure is made from yew, always associated with death because it is poisonous in all its parts. He is thin and wasted, his clothes in rags, and his bells are lead. He is plunging towards the well and Fortuna is about to cut his thread.

Walnut, lime, boxwood, mulberry, olive, lignum vitae, gold, silver, lead and semi-precious stones
28 inches tall

Lord of Misrule

"Then [Shakespeare's time] costumed fools still rolled and tumbled in the street and revelled in their obscenities." E. Welsford, *The Fool and his Sceptre*.

In medieval Europe, Christmas lasted 12 days, the last day being the Feast of Fools. This was a direct descendent of the Roman Saturnalia. On this day the world was turned upside down. Masters waited on servants, bishops on priests, and civil disorder was rampant. Wild festivals took place, and all was led by the Lord of Misrule, who was chosen, often because he was actually insane. In France the Lord of Misrule was *Mere Folle* (Mother Folly), and the goings on, particularly involving the clergy, reached such levels of depravity that inevitably the church and the law put a stop to the whole thing. This small carving shows a fat, drunken jester, mocking the world just before he falls over backwards.

This piece was unfortunately stolen from The White Knight Gallery in a smash and grab raid and has disappeared completely.

Walnut and gold
8 inches tall

Harlequins

Harlequin is one of the oldest characters in the history of theatre, traceable back to ancient Rome and beyond, yet still alive and well, as any visitor to the Venice Carnival will see. Of course, his character has changed somewhat from the lewd satyr of classical times, through the buffoon Arlecchino of the Italian comedy, the supposedly invisible sprite of the English Harlequinade in the eighteenth century, to the clown of more modern times. Harlequin has often been used as a vehicle for illustrating emotions that would seem alien to his comical nature, notably in Picasso's paintings of Harlequin's family.

In the Italian comedy he was dressed in a suit made of patches of coloured cloth, wore an ugly black mask and a small round hat, and carried a cudgel. Later this outfit developed into the familiar suit of diamonds and cocked hat, and Harlequin became a rather elegant, athletic figure. My harlequins are of this type, made from a solid wood figure inlaid with diamonds of different timbers, metals, shells and stones. They are extremely difficult and laborious to make.

Harlequin Abandoned

This sculpture was commissioned to celebrate the fiftieth birthday, on August 3 (Leo), of a lady who worked as a theatre and television costume designer. It depicts Harlequin as a clown in the theatre, bereft of his dresser, reduced to darning holes in his tights, in typical male fashion.

The main body of the figure is lime. It is inlaid with blocks of 50 different woods (for 50 years) in the shape of diamonds, traditionally associated, like gold and the sun, with the month of August.

The base consists of four stages. On each stage is a scene representing periods from the lady's life and arranged as the four seasons in reverse order. The cushion and the box are walnut on a marble turntable.

The book, which Harlequin stands on in disgust at being left by his dresser, is the script of *Miss Morrison's Ghost* for which the lady was nominated for a British Academy of Film and Television Award for best costume design, and during which period she met her husband who stole her away to eventually make her his wife. It is made of rhododendron from the garden of their home.

The buttons on the cuffs are gold, the metal of the sun, August and the House of the Lion. The buttons down the chest are carnelian and lapis lazuli. Carnelian is the month stone for August (conjugal felicity, success in legal matters and protection from scorpion stings), and lapis lazuli is the star stone for Leo (cure for apoplexy, epression and melancholy). The needle and thread are copper and silver, associated with the moon goddess, Diana, protector of women.

Detail, hat: Harlequin's hat is purpleheart, decorated with a poppy (the lady's favourite flower), made from coral (the name of her pet dog) and olivine (the birthstone for August believed to be protection against melancholy and the terrors of the night).

Detail, closeup, diamonds: Apart from 50 different species of wood, the figure is also inlaid with copper - associated with Venus, femininity, beauty and love; and lead – associated with Saturn and the golden age of retirement and pleasure.

Lime, walnut,
purpleheart and
mixed materials
20 inches tall

Side one, Winter: Her parents and brother, the two houses in which she was raised in the city of Crewe, the centre of the railway industry.

Side two, Autumn: Her husband, their first house and the dog Coral.

Side three, Summer: Her two children, two more dogs, their second house and her husband's film company logo holding the Palm D'Or, which he was awarded this same year.

Side four, Spring: The third house, the lady herself and symbols of her new profession as interior designer.

Harlequin, the Storyteller

*H*arlequin the Storyteller was commissioned by a famous British furniture maker to celebrate his family.

The Harlequin is a portrait of the owner. It is made of lime inlaid with various timbers, and the hat and shoes are walnut, his favourite wood. The opal buttons are the stones for the month of October, when he was born. The turquoise is for a child that died in infancy. As a Libran, the owner is under the rule of Venus, which is influenced by copper, in the ribbons and inlay, and amethysts, the stones in the hat.

The books he is sitting on are walnut, elm, yew, sycamore, rosewood and pitch pine. The titles are favourites of his and his three children, Tamsen, Heidi and Alice. The book he is holding is boxwood, and the quotation, "That first dawn in Helen's arms…" is from W. B. Yeats's "Lullaby"— Helen being his wife.

Alice is shown holding her companion Hippity. As a Capricorn, Alice is under the rule of Saturn, symbolised by her necklace of lead and turquoise. Heidi, a Libran, wears opal, copper and amethyst like her father. Tamsen was born in May, which is also under the rule of Venus, symbolised by amethysts and copper, but she also has an emerald for the month of May.

Lime and mixed media
16 inches tall

Harlequin the Politician

Harlequin the Politician is a portrait of British Prime Minister, Tony Blair. It was carved in 2000 when Blair was very popular; of course by the time this book is published, he may be history.

Blair was very popular and had a massive majority in Parliament. He invented "New Labour" and brought the language of the M.B.A. into politics. However, by 2000 it was clear that his ties with the old, left wing, working class, Labour party were getting very thin indeed. Unions were withdrawing their support and things were not going well. It was at this critical point that Blair is depicted. He had tried to be all things to all men. Symbolically he had painted himself the colours of every political party and, in doing so, has transformed himself into the clown Harlequin.

The whole figure is presented in a way rather reminiscent of a Victorian monument standing on a marble plinth. In this case, however, the marble, typically, is fake—just painted. It is all image. The four gargoyles supporting "Harlequin" are Mandleson, Brown, Prescott and, of course, his wife, Cherie.

The funny thing with this carving is that as time goes by Tony seems to grow more like the carving—like the protagonist does in *The Picture of Dorian Gray.*

Detail: Peering down through his newly acquired oval wire spectacles (oval ones show sensitivity!) with his mask lifted (a rare thing for a politician), Blair is busily painting himself bright red in a frantic attempt to regain the socialist heartland. In the base are four windows with painted heads set between the mirrors.

Lime and mixed media
40 inches tall

Side one: At the front is the eternally smiling Blair.

Side three: Blair out of control with his horns and fangs.

Side two: Blair the control freak looking rather manic.

Side four: A skull—yesterday's man.

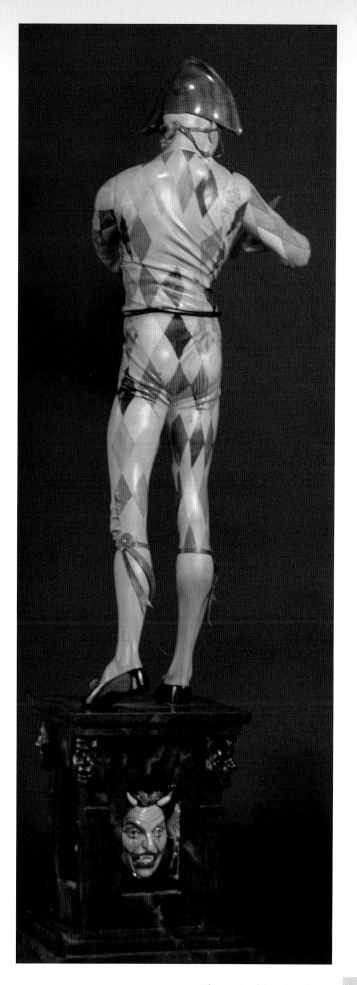

The Dream Maker

Dreams have always occupied a highly influential position in the human condition. In biblical times, and no doubt long before, dreams seem to have been regarded as a kind of communication with the Deity, foretelling the future, explaining the present and reliving the past. This may seem to be foolish superstition today, but Freud's ideas are pretty bizarre and the modern scientific explanations are not very convincing. Dreaming is often referred to as a pleasant experience, but personally I would describe myself as suffering from dreams, rather as one might suffer from allergies or a recurring illness.

In my carving, *Harlequin The Dream Maker* is a handsome, attractive figure. The only harlequin on a basic figure of walnut that I've made, this carving features diamond inlays made of materials that flash and glitter: mother of pearl, silver, gold, pau shell, oyster shell and various rippled timbers. His collar is polished brass and pink ivory wood. He is charismatic and glamorous, attracting us with the promise of pleasure, with his dream catcher to gather pleasant fantasies for us.

But the large embroidered eye should warn us that once we are asleep, that inner eye in our mind will be wide open, and we will be trapped inside The Dream Maker's world—the infinity cabinet—and be totally in his control like the anonymous dangling manikin. The endless vista of the cabinet, disappearing into darkness, reflects the capacity for dreams of endlessly repeatable horrors that we will be helpless to bring to an end. The pleasant, fluffy white clouds we should be floating on will transform into demons that lurk in the recesses of our minds. Sweet dreams!

Lime, walnut and mixed media; infinity model by Andrew Varah; embroidered eye by Ruth Burbery
36 inches tall

Corruption

I am not one of those dinosaurs who proudly proclaim their total and happy ignorance of computers. I am heavily into digital cameras, videos and producing CDs for computer use. However, I have never needed to use a computer, because my son or my wife has always done it for me. I cannot even switch one on. What I do know is the hours of frustration and irritation resulting from hardware or software not functioning properly. I am truly amazed when my son buys an expensive piece of software that actually does not work because it has known faults in it that must be rectified by more time spent on the Internet downloading a "patch." It is like buying a car with one wheel missing. I have also seen the problems of viruses getting into computers and the resulting damage.

Corruption came out of this half-baked knowledge of Internet technology. Harlequin, in his later manifestation, was a naïve and rather ineffectual servant. In this character, the Harlequin in Corruption symbolises the computer itself, which is trying to dispose of a corrupted hard drive. Unfortunately, he is too late and the corruption, like a contagious disease, has transferred itself to his hand, which is being rapidly rendered into pixels.

Lime, walnut, and various timbers
18 inches tall

Autumn Leaves

Not many people are happy about growing old, watching the aging of their bodies and the decline of their abilities. The realisation of one's mortality is a creeping awareness incomprehensible to the young. The sad-looking Harlequin in this piece has learned the bitter truth. He can no longer be the acrobatic buffoon of his youth. His suit of diamonds is disintegrating and blowing away like dead leaves. In his fingers he holds a petal from the dying flower in his hat. Strange clouds float around his head like amorphous memories of better times, and he looks out at us with a sad appeal.

Behind him is a Venetian colonnade, as seen in a convex mirror. Through the arches is a view of San Giorgio Maggiore across the Canale di San Marco with some gondolas riding at anchor in the foreground, waiting to ferry Harlequin across to the other side.

"Now it is autumn and the falling leaves
And the long journey towards oblivion…
Have you built your ship of death, O have you?
O build your ship of death, for you will need it."

—D. H. Lawrence

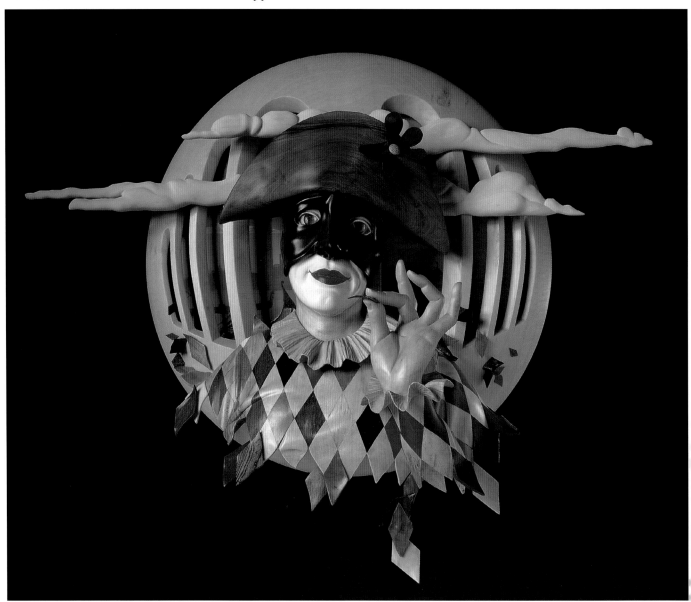

Lime, walnut, African blackwood and mixed media
18 inches tall

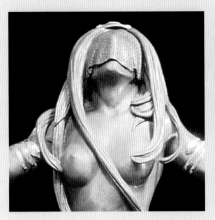

Mythology

Mythology can be treated in a superficial way, simply as fairy stories or as a primitive people's interpretation of life and the universe around them. Whilst one could not call the ancient Greeks and Romans primitive people and whilst what we call myths were their deeply held religions, the stories and beliefs that surround their Gods seem to reach back further and further into the far reaches of pre-history and encapsulate the most basic, atavistic urges of man.

I have made many carvings based on mythology. These are a few of them, treated fairly superficially I fear.

Gaea

In Greek creation mythology, the first two principals to emerge from Chaos were Uranus, the sky, and Gaea, the earth. The female principal, Gaea, has been worshipped from the remotest times as the Earth Mother.

From the union of Uranus and Gaea there appeared the Titans, the Cyclopes and three monsters. Uranus locked his offspring away in the depths of the earth, but Gaea made a sickle of iron and persuaded Cronos, her last-born, to castrate his father. Various terrible beings sprang from the blood, but the genitals, which Cronos threw in the sea, turned into white foam from which the Goddess Aphrodite was born near the island of Cyprus.

The cult of Gaea, the Earth Mother, persisted throughout the ages under different names—Rhea, Cybele, Demeter and others—but eventually disappeared under the weight of patriarchal religions.

Although ancient mythology may seem remote and bizarre, I believe that it was the way of explaining the world when the scientific explanations that we believe today did not exist, rather similar to the way children are given unlikely explanations for thunder – God moving the furniture, God bowling....

However, I think this belief that all natural phenomena were inhabited by a God or spirit engendered a reverence for the whole of creation that is sadly lacking today. Within living memory, lumberjacks in remote areas of Europe said a prayer begging forgiveness before felling a tree. Now, the Earth is there to be raped and pillaged for as long as it lasts. Thus, in my sculpture Gaea lies in her death throes over a blackened earth.

Someone today asked me if there is no ray of hope, but I think the story of the Earth is following the usual, inevitable course of a Greek tragedy. However, there is one possibility. Man was destroyed by the flood sent by Zeus because Prometheus had given them fire. Prometheus warned his son Deucalian who built an ark and survived. After the flood abated, Deucalian was told by the Goddess Themis to "cast behind you the bones of your ancestors." Deucalian obeyed and threw behind him rocks torn from the earth, the very bones of Gaea. The rocks were transformed into men and women and re-peopled the Earth. Perhaps the Earth can re-generate itself from the bones of Gaea.

Lime and marble
18 inches tall

Aphrodite

Like all mythological figures Aphrodite was a development from other Goddesses going back into pre-history. She had many forms and variations ranging from a fertility Goddess embracing all of nature through to the patroness of prostitutes. Such was her beauty and the power of love she imparted that the myths surrounding her are endless. Most famously, the Trojan Wars were started by Helen being given to Paris in return for him awarding Aphrodite the golden apple, signifying that she was the most beautiful of all beings.

The greatest sanctuary of Aphrodite was at Paphos in Cyprus where she was born of the sea foam, probably a story created from a pun on her name which means sea foam. She is often depicted standing in a shell as I have done here.

Of course, most depictions of Aphrodite are, like mine, merely an excuse to carve a beautiful woman.

Boxwood, gold, coral and marble
16 inches tall

Andromeda

Andromeda was the daughter of King Cepheus of Ethiopia. Cassiopeia, the queen, had offended the ocean nymphs, and Poseidon, the God of the seas, sent a sea monster to devour men and beasts. The oracle at Ammon decreed that only the sacrifice of Cepheus's daughter Andromeda to the monster could save the country. She was duly chained to a rock on the seashore until rescued by the Greek hero Perseus.

This version of Andromeda, my second, is based loosely on a painting by the Pre-Raphaelite artist Edward Burne-Jones and shows a young girl bound with ribbons. This is my own portrayal of the modern western woman, channelled by society into being a liberated "person," competing with men on their own terms and so on, but she will always be bound by her own femininity, represented by the ribbons, waiting for her Perseus.

Moros

"Above the Gods and above Zeus himself hovered a supreme power to whom all were subject: Moros, or Destiny. Son of the Night, Moros, invisible and dark like his mother, prepared his decrees in the shadows, and extended his inescapable dominion over all. Zeus himself could not set aside his decisions and had to submit to them like the humblest mortal. He had, moreover, no desire to set aside the decisions of Destiny; for, being himself Supreme Wisdom, he was not unaware that in upsetting the destined course of events he would introduce confusion into the universe it was his mission to govern. Thus even when it was a matter of saving the life of his own son, Sarpedon, the hour of whose death the Fates had marked down, Zeus preferred to bow his head and let what was ordained be fulfilled." F. Guirand, *The New Larousse Encyclopaedia of Mythology*

Lime, copper and marble
30 inches tall

Walnut, lime & marble
18 inches tall

Bride

Once upon a time, long ago, most people believed God was a woman, under various names and guises but basically the Earth Mother. The serpent was not always considered a symbol of evil but of intelligence, and sexual potency—as for instance on the symbol of the medical profession. However, a patriarchal religion wished to disassociate itself from this matriarchal heritage, so it turned the serpent into a symbol of evil that seduced the woman into committing the original sin, and both were condemned for many centuries to come. The patriarchal religion flourished in various forms, and the Earth Mother was consigned to the realms of folklore.

In this carving, The Bride, symbolising the mother Goddess, is veiled in readiness for her marriage with the serpent. In her outstretched hands she holds two spheres—one of golden quartz, representing the sun, and the other of labradorite for the moon.

Walnut, silver, golden quartz, labradorite and ruby
18 inches tall

Merlin

I am not sure whether Merlin is classified under mythology, folklore or history, but certainly he has appeared in many forms for a very long time. His association with the legends of King Arthur is almost purely Hollywood. In recent years Merlin has been portrayed as a rather demonic character controlling people and objects with his mind. I grew up with the idea of Merlin more as an alchemist making magic spells and potions. This is the type of figure I have depicted.

Merlin is seated at his ancient carved table holding a skull and studying a huge book, which is inscribed with magical symbols. On the tripod before him, liquid drips onto a toad that is metamorphosing into a man. The cat leaps onto Merlin's back in horror. The raven under the table, unperturbed, carries on pulling a worm out from between the flagstones of the floor. Around the base is a quotation from Marlowe's *Dr. Faustus*, an incantation for summoning Satan, "Orientis Princeps Belzebub, Inferni Ardentis Monarcha, Propitiamus Vos."

This was an extremely complex and difficult carving made from a large, solid block of lime. The base has been subsequently changed to one of oak fitted with a secret compartment to contain the drawings and paperwork relating to the piece.

The figure on the following page is my second version of Merlin. It is more of an Arthurian figure, complete with the sword Excalibur. The space inside the roots is filled with a clear resin encapsulating a representation of Merlin, symbolising his end, locked in a cave of ice.

Lime and oak
12 inches tall

Merlin II

Walnut, silver, bone and resin
14 inches tall

Literature

I have carved a great many figures based on characters in books; for example, over twenty from *Alice in Wonderland* and *Through the Looking Glass* and many from Shakespeare. Of course, there is an element of "standing on the shoulders of giants" about doing this, but I am certainly not the first artist to do it, and I will not be the last. I find it a fascinating challenge to try to encapsulate an archetypal figure, such as Puck or Falstaff, in one small piece of wood, and the greatest pleasure is when my vision coincides with that of my audience.

Ariel

Ariel is the airy spirit that carries out the bidding of Prospero in Shakespeare's *The Tempest*. I have tried to depict him as a rather demonic being, powerful and enigmatic, who is about to soar into the ether.

> "All hail, great master: grave sir, hail! I come
>
> To answer thy best pleasure; be't to fly,
>
> To swim, to dive into the fire, to ride
>
> On the curl'd clouds: to thy strong bidding task
>
> Ariel, and all his quality." *The Tempest*, Act 1, Scene 2

Under Prospero's orders he wrecks a ship in the first Act by appearing as fire that was brighter than lightning that

> "… most mighty Neptune / Seem to besiege and make his bold wave tremble / Yes, his trident shake."

Walnut, copper, gold, silver and semi-precious stones
16 inches tall

Falstaff

Sir John Falstaff is one of Shakespeare's most famous characters—people are still described as "Falstaffian"—and has provided many actors with some of their funniest roles. Although the Falstaff in the *Merry Wives of Windsor* may be a less solid figure than the one in the historical plays, he is still the same, cowardly, self-serving, lovable character.

My piece is based on that part of the *Merry Wives of Windsor* where Falstaff is in danger of being caught in *flagrante delicto* in Mistress Ford's bedroom. When this happened on a previous occasion, he was stuffed into a laundry basket and smuggled out. Now we see him desperately pulling up his hose and saying, "No I'll come no more in the basket." This was the second carving I made of Falstaff; the first one, several years previously, is featured in my book *Techniques of Creative Woodcarving*.

Walnut
16 inches tall

A Midsummer Night's Dream

I find this a fascinating play, which is a constant source of ideas, and I have carved a number of its characters.

This piece is the largest carving I have made, being six feet long, three feet high and seven inches thick at its deepest point. Intended to hang above the head of a bed, it is lit by two wall lights, the one on the left being the setting sun and the one on the right, the rising sun. In the centre the moon is also illuminated. Many of the figures in the carving are taken from the paintings of the Swiss artist Fussli.

Detail, left: The left side shows the artisans looking for Bottom, who is in the centre with Titania, The Queen of the Fairies, and some of her attendants. The figure in dark red is the owner of the piece, and his house features in the background.

Detail, right: The right side shows the dark figure of Oberon lurking in the background behind the musician and Puck flying through the trees, while the happy lovers are cavorting in the shrubbery.

Walnut and mixed media
36 inches tall

Puck

Puck is the impish fairy in *A Midsummer Night's Dream* who causes so much trouble with his antics. He is the servant of Oberon, the King of the Fairies, but his cynical humour and mischievous tricks have made him a popular figure—coining the description "puckish" for a certain type of person. Apparently the figure of Puck, or Robin Goodfellow, goes back into the mistier realms of folklore.

Amongst his other activities in the play, he is ordered by Oberon to squeeze the juice of the flower Love in Idleness, a purple pansy, onto the eyelids of Titania. This will cause her to fall in love with the first creature she sees on waking. This, of course, turns out to be the yokel Bottom whom Puck has endowed with an ass's head, thus creating one of the most bizarre scenes in literature.

My carving shows Puck holding the flower, with his fingers to his lips as if sharing with us his proposed mischief and instructing us to silence.

Briarroot burr, copper, ammonite and elm
8 inches tall

Ozymandias

I carved this from the root of a mesquite tree when I was on holiday in Texas. It is one of my few unplanned carvings, just done on the spur of the moment. I do not know what the details on the back signify; they are just doodles in wood. I have known the poem by heart since I was at school.

"I met a traveller from an antique land
Who said, 'Two vast and trunkless legs of stone
Stand in the desert. Near them, on the sand,
Half sunk, a shattered visage lies, whose frown,
And wrinkled lip, and sneer of cold command,
Tell that its sculptor well those passions read,
Which yet survive, stamped on these lifeless things,
The hand that mocked them and the heart that fed.
And on the pedestal these words appear:
"My name is Ozymandias, King of Kings:
Look upon my works, ye mighty and despair!"
Nothing beside remains. Round the decay
Of that colossal wreck, boundless and bare
The lone and level sands stretch far away.'"

—Percy Bysshe Shelley

Mesquite
15 inches tall

Sick Rose

"O Rose, thou art sick!
The invisible worm,
That flies in the night,
In the howling storm,
Has found out thy bed
Of crimson joy;
And his dark secret love
Does thy life destroy."
—William Blake

This poem by Blake is said to have concerned a woman friend who was infected with venereal disease. On the carving, the vine, which is wrapping itself around the woman, represents the insidious progress of this disease. The vine is made of yew, the wood of death, and its exotic flowers (roses) are red padauk and ebony. Despite this, she still has a look of post-coital satisfaction on her face.

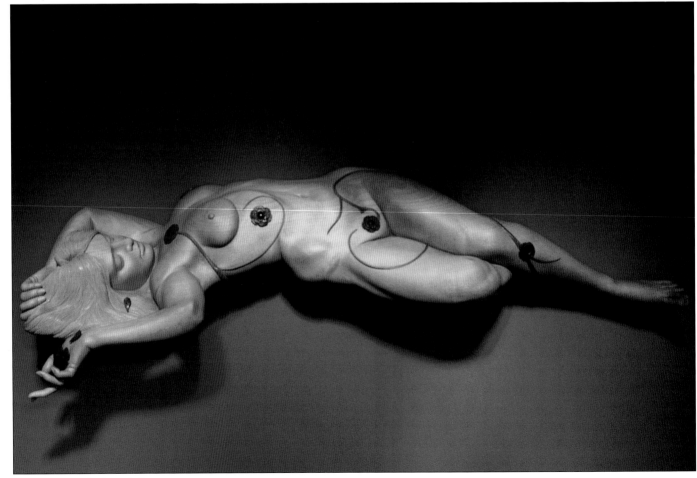

Lime, yew, padauk and ebony
36 inches long

Titania

In *A Midsummer Night's Dream*, Titania, the Queen of the Fairies, is locked in a dispute with her jealous and spiteful consort, Oberon. To achieve his ends, Oberon orders Puck, his scheming servant, to drug his wife by squeezing the juices of the flower Love in Idleness, a purple pansy, onto her eyelids. The effect of this drug is that Titania will fall in love with the first creature she sees on waking—as it turns out, Bottom, who has been given a donkey's head, also courtesy of Puck.

The size of the fairies in the play is very ambiguous—at one moment they appear small; at another, of human dimension. I have portrayed Titania as tiny, delicate and beautiful, with fragile butterfly wings, a tiny crown and long, exotic hair swirling around her body. She is standing on the purple pansy, the deep rich colour of purpleheart making a violent contrast with the purity of the flawless lime from which she is carved. Delicately holding her wand, she appears to have just landed on the flower petal, as light as thistledown.

Lime, purpleheart, pink ivory wood,
gold, silver and soapstone
15 inches tall

Sheba

The erotic poem, "The Song of Solomon," is part of the Bible that is not often referred to. I have heard its inclusion explained by the reason that the Bible celebrates all of God's creation, including sexual love. Despite this neglect, the Queen of Sheba, supposedly the subject of the poem, is practically a household name and has been featured in music as well as paintings over the centuries.

The Queen of Sheba visited King Solomon in Israel and apparently had a romantic encounter with him. Clearly, she was black. "…I am black, but comely, Oh ye daughters of Jerusalem, as the tents of Ké-där, as the curtains of Solomon," Old Testament, Song of Solomon, Chapter 1, Verse 5. The land of Sheba was supposedly somewhere around the horn of Africa. My portrait is of a rather savage-looking woman based on the ethnic type that now inhabits that part of the world.

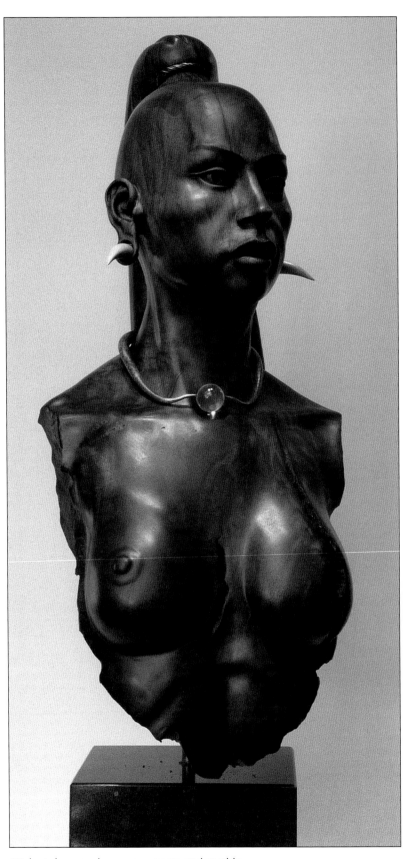

Walnut, boxwood, copper, quartz and marble
30 inches tall

Pierrot

Pierrot is a character originating in the French pantomime, representing a full-grown man with a child's mind, frequently drunk. The tallest, thinnest man was chosen to play the part, dressed in a white gown with very long sleeves and large buttons down the front, and covered his face in white powder. Pierrot still survives as a traditional circus clown.

Walnut, ebony and sycamore
20 inches tall

Death of Pierrot

This carving is based on the work of the artist Aubrey Beardsley. Beardsley, considered one of Britain's finest illustrators, died of tuberculosis in 1898 at the age of 26.

One of his last drawings shows Pierrot, a self-portrait, on his deathbed being paid a last visit by his friends. This triptych was made in 1998 as a tribute to Beardsley, and all of the figures are adapted from his drawings. The paintings on the outside of the closed triptych are his illustrations of John the Baptist and Salome.

When open, it reveals a Venetian street scene during the Carnival, with Beardsley's bizarre figures milling around amid dozens of coloured balloons. Pierrot can be seen through a small window, lying in bed, behind the left-hand figure of the central panel; the great artist is dying, but his creations live on.

Walnut, lime, copper
and acrylics
48 inches tall

Fantasies

Although I have called this section "Fantasies," clearly some of the carvings in other parts of the book could also be in this section. One might say that even a simple nude is a fantasy.

I suppose my own feeling is that a fantasy should not really need a title, but should invite a suspension of reality by the viewer who can then immerse himself in the work of art and interact with it and, in fact, extend it. This would be true of many of Dali's paintings where the titles and content are so obscure as to be meaningless to most of us who, nevertheless, feel deeply involved with his pictures. Few of my pieces reach this level of mystery, but I hope that most of them remain meaningful without an explanation.

The Alchemist

Alchemy was the genetic engineering, the rocket science of the medieval period, and it continued to be avidly pursued for several hundred years by people now considered advanced scientists, Sir Isaac Newton for example. Alchemists believed that all matter was created from the four elements of Air, Earth, Fire and Water combined in different mixtures, which formed the different substances. If you could change the mixture, you could change the substance (e.g., from lead to gold). This does not seem a million miles from modern science where elements can indeed be changed by removing or adding various atomic particles. However, the alchemist's methods of trying to achieve this end were sometimes a little exotic, to say the least.

Detail, head: The alchemical process is as much a psychological one as a chemical one and was written about extensively by Carl Jung. Here, the balls issuing from The Alchemist's brain represent his creative mental powers. The face is based on a drawing by Earnst Fuchs.

Detail, upper body: The mandrake is made of 3,000-year-old Irish bog oak, and the main figure is holly. Holly is normally very white or, when left to season, a rather dirty grey. The log from which this carving is made was bright orange, due, I imagine, to some fungal attack in the tree. The wood was perfectly sound And matured to a warm pale brown.

My carving shows the alchemist emerging from a green bubbling miasma—the prima material—surrounded by hundreds of balls which spiral around him, rather like DNA, issuing from his body and his brain. Many of the balls are held on fine copper threads connected to the alchemist's eyes and fingers. Suspended before him is a mandrake. The mandrake root was, like ginseng, often of a human shape, and like ginseng, the mandrake root was considered magical. However, when a mandrake was uprooted, it emitted a scream, which was fatal. The alchemist therefore enlisted a starving dog, tied it to a mandrake, left a piece of meat nearby and went away. The dog then pulled up the mandrake and was killed by the scream. The alchemist could then perform certain operations on the mandrake, which turned it into a homunculus—a small man who had magical abilities and was under control of the alchemist.

Holly, copper and bog oak
24 inches tall

The Astrologer

As a non-believer, astrology appears to me to be one of the great manifestations of human gullibility and irrationality that it is possible to conceive. This global industry has for centuries extracted money from people in every station of society with not the remotest possibility of getting anything for their pains except delusions. But perhaps I am wrong and tomorrow one-twelfth of the world's population will have a bad day because of the proximity of Saturn to Leo, or whatever.

My astrologer is wearing the hooded cloak of a monk—this gives him the authority of the church. However, the elegant, well-shod leg protruding from the cloak indicates someone other than a priest underneath. The mask he wears also symbolises duplicity. His cloak is belted with a rope, and tied up in the rope is a snake swallowing its tail. This is Ourobouros, the serpent of eternity.

It is implied that the astrologer has control over destiny, and this is supported by the book he holds—the Liber Mundi—in which is inscribed the fate of everyone in the world. Carved in his cloak are several eyes suggesting his all-seeing abilities. In the other hand he holds a tetrahedron made of Egyptian obsidian, for all the "science" of astrology originates in Egypt.

Thus the astrologer stands and makes his pronouncements, but sneaking around from behind him is an ugly dwarf blowing soap bubbles. The base is inlaid with the seal of Solomon on the top, and around the sides are painted the star formations of the Zodiac.

Detail: Traditionally the dwarf represented man's inability to control nature and life. He mocks the astrologer by blowing bubbles symbolizing his empty words. The quartz spheres, representing soap bubbles, symbolize the empty, unrealistic pronouncements of The Astrologer. The Zodiac was drawn up two thousand years ago and does not relate to the actual constellations in the sky because the earth's axis has shifted; for example, the sun's movement through Scorpio now actually only lasts seven days.

Walnut, olive, lime, yew, padauk, cocobolo, quartz, obsidian, copper and mixed media
24 inches tall

The Dagger

In 1634, Jeanne de Anges, the Mother Superior of the Ursuline nuns of Loudon in France, and a group of very young nuns became sexually obsessed with their parish priest, the Curé Urbain Grandier. Whatever part Grandier may have played, he ended up being burnt at the stake as a witch, convicted of using satanic arts to seduce the nuns. Although the nuns are now believed to have been suffering from lack of sunshine, starvation and various other physical ailments, several books and films have been produced about the episode.

My carving shows a symbolic priest wearing a mask half black, and half white to show the good and evil sides of his nature. On the white side is a nun, made from flawless lime. She grasps the blade of the dagger, a symbol of the hard, religious life and points accusingly at the other woman. This other woman, on the black side, is no longer dressed as a nun and is made from diseased, spalted lime. She grasps the handle of the dagger, with its phallic shape, showing her fall from grace.

Walnut, lime, sycamore, ebony, copper and tiger's eye
24 inches square

The Drummer

In the medieval period a series of plagues swept across Europe, killing some estimated one-third of the population. Entire towns and villages were wiped out as the hideous deaths by bubonic plague were followed by starvation, other diseases and the inevitable, never-ending wars. Many were convinced that this was a divine punishment and heralded the end of the world. Bizarre religious sects appeared like mushrooms, such as the flagellants, who drifted round the countryside whipping themselves.

Amongst the strange superstitions that arose was one which believed that the dead rose from their graves in the dark of the night and played music to entice others to join them in the grave. This "Dance of Death" was, of course, apparently successful, as every night more victims succumbed to the plague.

The Drummer, blindfolded to show his indiscriminate appetite, manically beats his drum, laughing insanely and standing on his own tomb—a stark reminder of his own mortality.

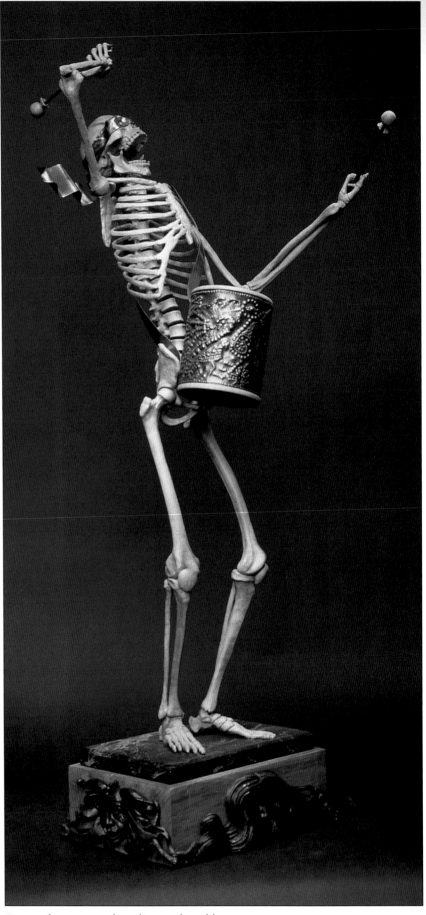

Boxwood, copper, walnut, lime and marble
36 inches tall

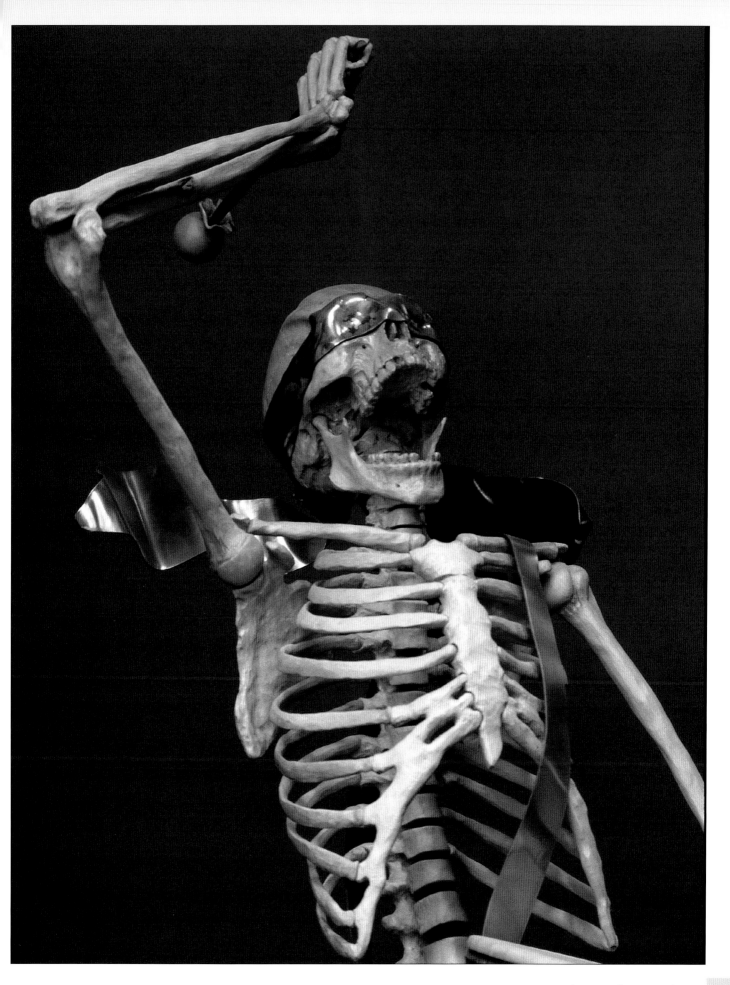

Theatre Mask

I have carved many masks. This theatre mask is made from lime and walnut inlaid with ebony and red padauk.

Walnut, ebony, recycled ivory, amber, silver and brass
15 inches long

Venice

At first sight the fascinating city of Venice seems to be a kind of theme park for tourists, where everything has been contrived to make it more and more quirky, curious and unreal. Three miles across the bridge from the industrial wasteland of Mestre you enter a world where everything is done in boats or on foot. Every brick, bottle of beer or packet of razor blades has been brought on a barge and then carried on a sack truck to the shop or bar. But Venice is not a theme park—it is a working city, full of workshops and artisans making beautiful objects, much of it for the tourists.

In late February the Carnival begins. For no apparent reason, tourists who wandered the streets like normal people put on bizarre masks, funny hats, and elaborate capes and costumes. When the carnival finishes, they all disappear—everyone becomes normal again.

The three figures in *Venice* are typical. They may look sinister and exotic, but underneath they are simply German bank clerks on holiday.

Secrets

The Venice carnival was stopped by Napoleon and only reinstated in 1980. However, its origin goes back to the heydays of the Venetian Empire when great festivities were held in the city. People who wanted to indulge in these unseemly activities wore masks to conceal their identities—wealthy aristocrats visiting courtesans, priests and nuns out for a good time—all could hide behind their costumes and preserve their anonymity.

Secrets shows the view through a Venetian window looking out into the night. Crime must have been rampant in Venice because every single window, even up to the third and fourth floors, is heavily barred with massive iron grills. Looking out through the window, we see a huddle of masked figures planning who knows what assignation. With a heavily costumed figure it is almost impossible to tell anything about the person underneath, even their sex. Whatever they are planning, they seem to be being watched by another figure, perhaps one of old Venice's infamous secret police.

Walnut
30 inches long

The Temple

Lime, walnut and mixed media
29 inches tall

The building symbolises the human persona with its weak outer mask, behind which an endless array of other masks lurks in the convoluted passageways of the mind. On the top, the domes represent the higher aspirations floating away into the clouds; below, the decaying, atavistic, animal urges. On the left is the impenetrable citadel of Kafka's castle, on the right the equally unapproachable tower of spiritual fulfilment.

Europhobia

Detail, head: John Bull is shown here looking rather like his bulldog.

Europhobia is my comment on the xenophobic "Little England" mentality currently fermenting in certain sectors of British society. This tendency is personified in the figure of John Bull and his old bitch Maggie.

John Bull was created as a recruiting poster in the First World War and was re-created as Uncle Sam in the United States. John Bull became a symbol of British imperial power and aggression, as in the IRA song "Patriot Game"— "...six counties lie under John Bull's tyranny." The image persisted for many years, well into the 1960s, but has faded along with Britain's power. His cane handle is the head of an African, still failing to appreciate the civilizing influence of John Bull's grasping fingers.

He stands astride a square world showing the British Empire at its height. He would never have let "The Empire" go, and would rather die than give up sterling, yards, feet and inches, warm beer and overcooked food.

Detail, dog: Maggie, his dog, is clutching a tattered European flag, its gold stars peeling off to reveal the tricolour of Germany.

Lime, walnut, boxwood, yew, dyed veneer, pearl and amber
54 inches tall

Genesis

Bog oak is wood that has been in a peat bog for several thousand years. The effect of this is to make the wood very dark brown or jet black in colour and, when dried, extremely hard. In Ireland bog oak is found continuously in the process of cutting peat for fuel—it is a nuisance and is piled up at the sides of the cutting area. No one collects the wood for fuel anymore, although it burns well when dry.

I dragged this piece out of a bog in County Clare in Ireland (it is extremely heavy when wet) and took it home and left it to dry for a couple of years and then made Genesis. When dry, the bog oak has a grey, crusty surface, almost like stone. When this is removed, the wood underneath is very dense and black, like ebony, and, although very difficult to cut, takes a wonderful finish, as can be seen on the goat's head at the top. This really is the driving force behind this piece—the contrast between different surfaces and textures. That contrast is seen between the rough, jagged tree root and the polished smoothness of the serpent; between the dense, black oak and the delicate, white limewood baby; and of course metaphorically, between the ancient, black wood bringing forth new life.

Bog oak, walnut, lime, red padauk, elm, tiger's eye, amber and copper
44 inches tall

Detail, Goat's Head: The goat Amaltheia suckled the infant God Zeus, in gratitude for which services he made her into a constellation. From her hide he made the Aegis, the breastplate of the Goddess Athene. One of her horns was given to the nymphs conferring on it the marvellous property of refilling itself inexhaustibly with whatever was wished for – the cornucopia or 'Horn of Plenty'. Amaltheia looks down benevolently on Creation. On the other side the goat's head is reduced to a skull reminding us that the whole of Creation will one day die.

Detail, Old face: An ancient face, set with stones, appears to be dissolving into the wood.

Detail, Serpent: The serpent, exotic, beautiful and evil rears over the baby threatening the future.

Detail, Eggs: Eggs and spawn exude from cracks in the bog oak.

Tree Spirits

In pre-Christian times the inhabitants of the vast primeval forests of Europe believed that trees were inhabited by benevolent spirits whose help could be sought by banging on the tree trunk. This has survived to the present day in the form of "touching wood" or "knocking on wood" for good luck. When a tree was cut down, prayers had to be offered to appease the spirits in the tree, and failing to do so was a serious offence. Nowadays we cut down trees at a rate that has to be described, not in numbers or square miles, but by comparisons with countries of similar area to one year's logging.

Tree Spirits is a commentary on the ecological disaster we are perpetrating. It is made from a section of the trunk of a dead mulberry tree, covered in small burls.

Mulberry, quartz and pear
36 inches tall

Detail, ivy: Where the fabric of the tree is decayed, the spirit of the tree looks sadly out.

Detail, hands: On the top, an emaciated, dying hand reaches out to grasp a baby's hand. The only hope is that the next generation may be better than ours.

Detail, skull: The green man is an ancient fertility symbol found on buildings, including churches, all over Europe, and it did, in fact, migrate to America. Typically it features a face with leaves bursting forth from the mouth, ears, nose, eyes and the top of the head. I have depicted a dead green man, a skull sprouting dead twigs.

Detail, forest: In the heart of the tree, in a dense forest of trunks is a gleaming quartz wand, its very life source. Fungus is growing all over the tree. On the right is a fat grinning face—mankind, like a greedy schoolboy.

The Box of Delights

The theme of this carving is not intended to be of religious significance although it has some biblical content. It is intended more as a satirical, perhaps slightly cynical, view of the life cycle as it was propounded in bygone days by the moralizers of the Christian Church in order to keep the peasantry under their thumb. I flatter myself that I may have a little of the humour of the misericord carvers of the medieval period, if not their skill.

The sphere at the top represents the world held in a claw. Below it are two serpents entwined in a Celtic knot, one having a forked tongue, the other a single tongue representing good and evil. Evil, of course, will win out, and below the serpents are faces expressing the characters of the seven deadly sins, although here there are eight.

The final frieze around the base depicts Purgatory—no fire and brimstone, but a Never-Never-Land where the personality is destroyed, symbolized by a distorted head; individuality and choice are removed, symbolized by two bodies floating in space, and the joy and the richness of life cease to exist. Of the Jester, only the grinning skull remains. However, on the fourth side a girl with a secretive, mischievous smile holds out hope for those who have played their cards right. Even in hell there will be privileged classes.

The purpose of all this is, in fact, to decorate and conceal a secret draw inside the block which can only be opened by removing various hidden catches, which are to be found by following the clues in the text.

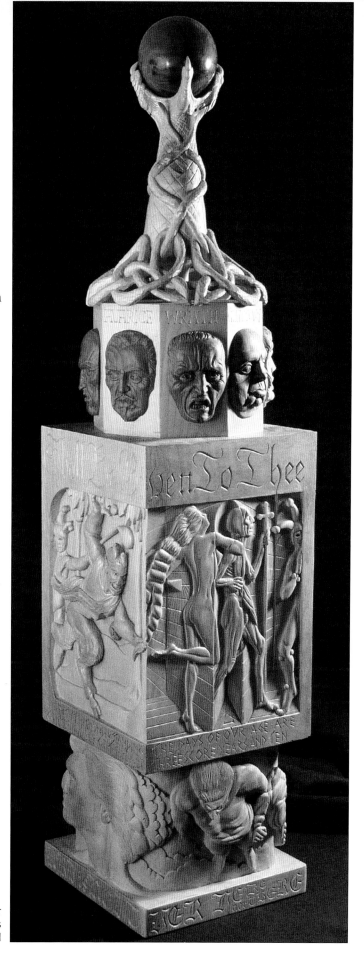

Lime, thuya root burr, silver
and various timbers
30 inches tall

Detail, Eve: Below the heads is an inscription, which reads, "What reward shall be given to thee false tongue." Below this are four panels, the first depicting Eve sitting under the tree holding an apple and looking a little pleased with herself. In the branches, the serpent, with the head of a jester, laughs at the storm that is about to break and at Eve's defence in the inscription below: "The serpent did beguile me and I did eat."

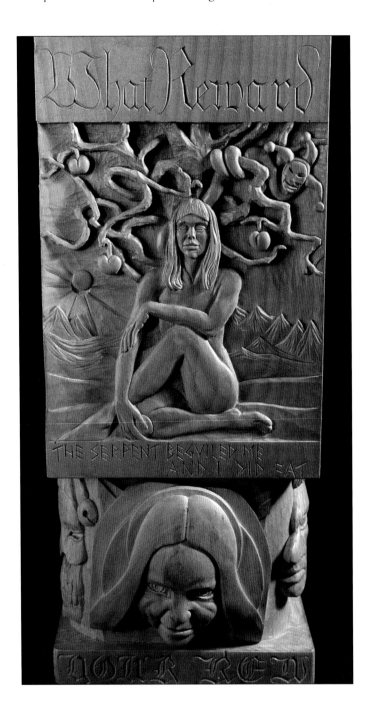

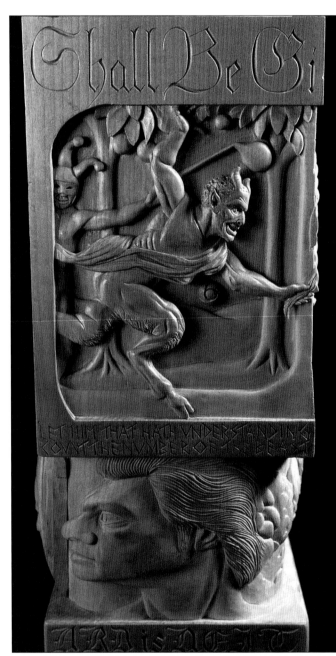

Detail, Devil: On the next panel, a rather jolly-looking Devil rushes through a wood in pursuit of a young female who has disappeared around the corner. The Jester peers from behind a tree, laughing at the Devil's efforts to corrupt the totally corrupt, and waves his bladder at him. A line from the Book of Revelations is carved under the panel: "Let him that has understanding count the number of the beast."

Detail, woman: The pursued lady is featured on the next side, running friskily into a doorway, from which she appears as a shrivelled old woman, who, being no longer desired for sin, has taken the cross. On the left, a sundial counts away the fleeting moments of youth, and on the right the Jester smirks at the newfound religion of the old hag: "The days of our age are three score years and ten" is inscribed below.

Detail, Death: Finally, on the fourth side King Death waits with open arms, an hour-glass in one hand and a sword in the other. Behind him, receding perspectives of emptiness vanish into a door-way. Like a rising sun, the Jester's head appears behind the door, mocking even death. Below this panel is written, "There was a door to which I found no key," from the Rubaiyat of Omar Khayyam. No key is needed to enter the House of Death.

Il Serenissima

Anyone who has visited Venice will have been struck not only by its unique beauty but also by the fact that most of it seems to be crumbling away. Of course it is common knowledge that the city is slowly sinking into the mud on which it is built, but the general neglect almost amounts to vandalism.

Il Serenissima, "the most serene," which is what Venice liked to call itself in the past, depicts a typical Carnival reveller with her face painted, appearing from a wall of crumbling brick and stucco. Through two archways, Venetian canal scenes carved in relief

can be viewed. The whole is carved in the form of a dome as if seen in a convex mirror.

The Carnival, wonderful though it is, sadly helps to destroy Venice by the sheer weight and volume of people and the increased traffic on the canals. It seems characteristic of Venice that its beauty is only a mask over a decaying reality underneath. Still the prophets of doom seem to have been foretelling the end of Venice as long as I can remember, and it is still there. In Canaletto's paintings it looked pretty flaky 250 years ago, when Venice was considered to be the most corrupt city in the world.

Lime, acrylics and gold leaf
14 inches in diameter

Animals

Wildlife seems to have an irresistible attraction for woodcarvers, which I also succumbed to. However, it soon became apparent that this was all rather illusory. Most animals and birds are virtually never seen without a great deal of time and effort being expended, and then only fleetingly. Of course, one can go to zoos, which are generally pretty awful places and have the same repertoire of exotic species with faded fur and stereotyped behaviour.

So when I wanted to study an ordinary British species, I would either have to spend half my life lurking in woods at night or use photographs, which is of course what happened. This began a process whereby the real animal became more and more remote as I carved a kind of symbol of a creature about which I knew almost nothing. It came as a shock when occasionally I saw a real animal and realised that this beautiful, sleek, living thing really had nothing in common with my lump of wood. Eventually I stopped carving wildlife.

Whales

The main reason I stopped making carvings of wildlife was the simple fact that I very rarely saw any and since I dislike zoos intensely, there was little likelihood that I ever would. Besides the wildlife population of England is mostly small furry rodents of limited appeal as sculpture.

However, there can be few more strangely exciting and moving moments than seeing a great whale surface alongside a small boat or the awesome power of a hump-back breaking. These are now hackneyed icons of environmentalist's attitudes, but all the photography and films in the world do not diminish the reality, and when I saw them in Hawaii and Iceland I did not have what you would call a ringside seat.

These whales are carved from a large block of European walnut and mounted on glass cullet. Cullet is the raw material of glass, straight from the furnace. Large pieces are very beautiful, full of refractions and faults, but glass cullet has become very difficult to acquire.

European walnut and glass cullet
18 inches tall

Dolphins

I first saw a dolphin in a zoo where it was the sole survivor, its fellows having died as the zoo was awaiting a new consignment. This was my last visit to a zoo. I saw dolphins again swimming alongside the ferry crossing from the North Island to the South Island in New Zealand. This was such a moving experience that I decided to carve them.

In this sculpture I have used a large dry root of a dead oak tree as a mount to convey the idea of these beautiful creatures swimming beneath the sea.

Walnut and oak
18 inches tall

The Monkey

This small carving, made in 1987, represents the first time I painted a piece. This was quite a decisive step, because it would have been extremely difficult to remove the paint if the result had been unsatisfactory. But in the end, everyone seemed to like it, and I have been painting carvings ever since. The eyes were inlaid with ebony, which gave a very good impression of the dark liquid eyes of the animal.

Walnut, ebony, gold,
amethyst, marble & oil paints
10 inches tall

Dominion

In 1989 my wife Betty, who founded and ran the British Woodcarvers Association, organised an exhibition to raise funds for the World Wildlife Fund. The exhibition consisted of many carvings of endangered wildlife created by the Association's members, which were displayed at 27 different venues, a percentage of sales going to the charity. To initiate the exhibition, a logo was needed for public relations, and Dominion was created for this purpose. It shows the dove held by a pair of hands, which may be protecting it or may be crushing it—the same option that Man has with the whole planet. The title Dominion derives from the verse in Genesis 1:26, "And God said, '…and let them have dominion over the fish of the sea, and over the fowl of the air, and over the cattle, and over all the earth, and over every creeping thing that creepeth upon the earth.'" The piece was featured on the cover of the exhibition catalogue, which also featured a foreword by H.R.H. Prince Philip.

The dove and hands, which are approximately life-size, are carved from limewood, which has been bleached paper white.

The piece was sold to Dr. Barbara Zeidler who bequeathed it to "Nature in Art," the British National collection of wildlife art, where it is now on display.

Lime
24 inches tall

Falcon

I carved several falcons and other birds of prey, mainly because there was a large falconry center nearby where one could observe these birds close up. I stopped when I became disenchanted with the fact that my birds were never going to fly, which is really what birds are all about.

Walnut
16 inches tall

Racehorse

Before I started woodcarving, I was a painter, specializing in racehorses, so it seemed a natural progression to carve them. However, in wood one thoroughbred looks much like another and I began to carve different types of horses and action scenes. This racehorse is one of a pair that fitted together running neck and neck.

Walnut
12 inches tall

Arkle

This carving is a portrait of the greatest steeplechaser of all time, whose astonishing victories made him a legend in his lifetime and caused a bronze statue to be erected at Cheltenham Racecourse, the scene of his greatest triumphs.

Walnut
12 inches tall

The Tiger

This large carving was my last attempt at animals. It seemed to me that a big cat without any coloring was pretty anonymous, so a tiger either had to be painted or textured in some way to create his stripes.

I actually saw this tiger the last time I ever went to a zoo. He is presumably still walking round and round his pen.

The entire tiger was textured to resemble hair and the stripes were cut in more deeply. Although this worked in certain lights, on the whole it did not.

Limewood
Size 30 inches long

The Making of a Masterpiece: Seraglio

The Seraglio is the Palace of the Sultans of Turkey at Constantinople (now Istanbul). It is situated on the Golden Horn and enclosed by walls more than seven miles long. The main building is the Harem, or "sacred spot," which contains numerous houses, one for each of the Sultan's wives and others for his concubines. I visited the Seraglio a few years ago but was unable to enter the Harem although its occupants are long gone.

Harems still exist in some countries. The horrors suffered by the girls, who have no rights whatsoever, to satisfy the lust of one man can only be imagined. Smoking a water-cooled hookah pipe, perhaps with a little hashish, must be a welcome escape.

18 inches diameter
Lime, walnut, copper
and acrylics

The work on Seraglio in Ian's studio in Ireland.

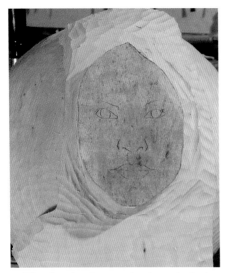

1. The drawing is traced onto a six-inch block of limewood; the convex background is roughed out.

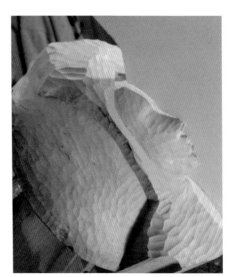

2. The face is roughly shaped.

3. More detail on the face and drapery is carved.

4. The face is completed and roughly sanded.

5. The face is completed and sanded smooth.

6. The hand is carved separately.

7. The arches are cut in, the hand is positioned, and the whole carving is sprayed with a primer.

8. The painting process is in its first stage.

9. The final touches, such as the detail on the background and the smoke, are finished.

The Early Years: Paggliacci (1975)

This small figure of a character from the Commedia dell Arte is the first carving I sold while still at art college. I had been carving in my spare time for about two years. Being short of money at Christmas I let it go for 30 pounds sterling (a low week's wage in today's money).

Plum
10 inches tall

The Early Years: The White Knight (1980)

Having spent a few years doing antique restoration, I decided to go into sculpture and opened the White Knight Gallery. It was so called after the character in Lewis Carroll's *Alice Through the Looking Glass*. This was my first major piece and is taken from Tenniel's illustration. It was also my first constructed piece, made from 126 pieces of lime.

Lime
24 inches tall

The Early Years: The Boast of Heraldry (1982)

This piece is one of several armoured figures that I made in this time period. It is based on a verse from Thomas Gray's *Elegy Written in a Country Churchyard.*

> The boast of heraldry, the pomp of power,
> And all that beauty, all that wealth e'er gave,
> Awaits alike th' inevitable hour:
> The paths of glory lead but to the grave.

Lime
36 inches tall

The Early Years: Elgin Horse (1987)

This head is a three-dimensional interpretation of a horse's head on the Elgin Marbles, the stone relief carvings from the frieze of the Parthenon in Athens, now displayed in the British Museum, London.

Maple
14 inches tall

The Early Years: The Pied Piper (1988)

A small, delicate piece based on Arthur Rackham's illustrations to the Browning poem.

Olivewood
14 inches tall

The Early Years: The Headsman (1989)

This rather sinister figure marked the beginning of a long series of masked figures. I feel masks are able to create considerable personality changes in people—making some capable of awful activities such as beheading prisoners—who then revert to a normal person when the mask is removed.

Black walnut
18 inches high

The Early Years: The Enchantress (1990)

A technically difficult piece made from a beautiful piece of wood with a mask of hammered silver set with jewels.

Walnut, silver and semi-precious stones
16 inches high

The Early Years: Colonel Blood

The Crown Jewels were the object of one of the most sensational robberies in history, when the deputy jewel keeper of Charles II's reign was battered on the head by Colonel Thomas Blood, disguised as a parson. Blood proceeded to treat the jewels as roughly as he had the keeper, banging the crown into shape to fit his container and made off with the crown, orb and scepter. When Blood was arrested, Charles seems to have admired the Colonel's audacity, for his Irish lands, which had been forfeited in 1660, were later restored to him and he became a gentleman of the Court.

When the Wall Walk of the Tower of London was opened to the public by Her Majesty the Queen Elizabeth of England, it included the Martin Tower, home of the Crown Jewels in Blood's time. Such is the fame of Blood that architects of the new exhibition felt that a significant memorial to him should be created in the form of a large carved portrait, which I had the honour to be commissioned to carry out.

Oak and lime
48 inches tall

More Great Project Books from Fox Chapel Publishing

Carving Classic Female Faces in Wood
By Ian Norbury
Learn to sculpt the female face in wood from renowned woodcarver and instructor, Ian Norbury. In this book the author not only teaches the fundamentals of woodcarving, but also demonstrates how to accurately and realistically portray the female face in wood. Clear, step-by-step photographs, complete with instructional captions, will guide you though an entire carving project. This is a must-have reference book for anyone interested in getting started or improving their realistic facial carving.
ISBN: 1-56523-102-3, 88 pages, soft cover, $17.95

Carving Classic Female Figures in Wood
By Ian Norbury
A fascinating gallery of carvings from one of the world's leading sculptors. This book is an opportunity for lovers of fine art to examine the work of one of the art form's living masters, Ian Norbury. Over 100 full-color photographs is included, plus commentary by the artist.
ISBN: 1-56523-221-6, 72 pages, soft cover, $17.95

Carving Eyes
By Jeff Phares
Life-like, expressive eyes are the key to successful human carvings and now you can learn how to create them with world-renowned carver, Jeff Phares. You'll learn to carve an average eye, a heavy-lidded eye, a baggy eye, a winking eye, and a sleeping eye. Includes over 200 full-color, step-by-step photographs.
ISBN: 1-56523- 163-5, 72 pages, soft cover, $14.95

Carving Ears and Hair
By Jeff Phares
Learn the art of carving authentic looking ears and hair with world-renowned carver, Jeff Phares. First, you will learn the secrets to carving realistic ears through photographs from several angles, drawings, and step-by-step instructions. Then, you will learn to carve curly and braided hair in a similar teaching style.
ISBN: 1-56523-164-3, 72 pages, soft cover, $14.95

Carving the Nose & Mouth
By Jeff Phares
Carve perfect noses and mouths with help from world-renowned carver, Jeff Phares. This book includes extensive step-by-step demonstrations, close-up photos and anatomy information on Native American and Caucasian faces.
ISBN: 1-56523-161-9, 72 pages, soft cover, $14.95

Carving Found Wood
By Vic Hood and Jack A. Williams
Inside this book you will meet some of our nation's most acclaimed artists that specialize in carving driftwood, burls, cypress knees and other forms of weathered wood. You'll learn their secrets to remarkable carving, be inspired by a stunning photo gallery of their work, and be led, step-by-step, through a cottonwood bark carving of a human face.
ISBN: 1-56523-159-7, 96 pages, soft cover, $19.95

Illustrated Guide to Carving Tree Bark
By Rick Jensen and Jack A. Williams
You will never look at a tree the same again! In this book, you will learn the specialized technique of carving figures in tree bark. Included is a complete guide to the various species of Cottonwood bark. A thorough step-by-step carving project of a magical tree house is included along with a beautiful gallery including woodspirits, animals, whimsical tree houses and much more.
ISBN: 1-56523-218-6, 80 pages, soft cover, $14.95

Caricature Carving from Head to Toe
By Dave Stetson
Find out what makes a carving "caricature" with this top-notch guide from Dave Stetson. First you will learn how anatomy relates to expression by creating a clay mold. Then, you will follow the author step-by-step through an entire carving project for an Old Man with Walking Stick. Additional patterns for alternate facial expressions, overview of wood selection, tools, and an expansive photo gallery is also included.
ISBN: 1-56523-121-x, 96 pages, soft cover, $19.95

CHECK WITH YOUR LOCAL BOOK OR WOODWORKING STORE
Or call 800-457-9112 • Visit www.FoxChapelPublishing.com